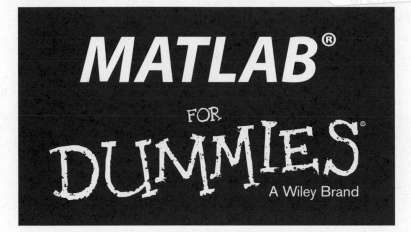

MATLAB®

FOR

DUMMIES®

A Wiley Brand

by Jim Sizemore
and John Paul Mueller

MATLAB® For Dummies®

Published by: **John Wiley & Sons, Inc.,** 111 River Street, Hoboken, NJ 07030-5774, www.wiley.com

Copyright © 2015 by John Wiley & Sons, Inc., Hoboken, New Jersey

Published simultaneously in Canada

For general information on our other products and services, please contact our Customer Care Department within the U.S. at 877-762-2974, outside the U.S. at 317-572-3993, or fax 317-572-4002. For technical support, please visit www.wiley.com/techsupport.

Wiley publishes in a variety of print and electronic formats and by print-on-demand. Some material included with standard print versions of this book may not be included in e-books or in print-on-demand. If this book refers to media such as a CD or DVD that is not included in the version you purchased, you may download this material at http://booksupport.wiley.com. For more information about Wiley products, visit www.wiley.com.

Library of Congress Control Number: 2014940494

ISBN: 978-1-118-82010-0 (pbk); ISBN 978-1-118-82003-2 (ebk); ISBN 978-1-118-82434-4 (ebk)

Manufactured in the United States of America

10 9 8 7 6 5 4 3

Contents at a Glance

Table of Contents

Introduction

● ●

MATLAB is an amazing product that helps you perform math-related tasks of all sorts using the same techniques that you'd use if you were performing the task manually (using pencil and paper, slide rule, or abacus if necessary, but more commonly using a calculator). However, MATLAB makes it possible to perform these tasks at a speed that only a computer can provide. In addition, using MATLAB reduces errors, streamlines many tasks, and makes you more efficient. However, MATLAB is also a big product that has a large number of tools and a significant number of features that you might never have used in the past. For example, instead of simply working with numbers, you now have the ability to plot them in a variety of ways that help you better communicate the significance of your data to other people. In order to get the most from MATLAB, you really need a book like *MATLAB For Dummies*.

About This Book

The main purpose of *MATLAB For Dummies* is to reduce the learning curve that is a natural part of using a product that offers as much as MATLAB does. When you first start MATLAB, you might become instantly overwhelmed by everything you see. This book helps you get past that stage and become productive quickly so that you can get back to performing amazing feats of math wizardry.

In addition, this book is designed to introduce you to techniques that you might not know about or even consider because you haven't been exposed to them before. For example, MATLAB provides a rich plotting environment that not only helps you communicate better, but also makes it possible to present numeric information in a manner that helps others see your perspective. Using scripts and functions will also reduce the work you do even further and this book shows you how to create custom code that you can use to customize the environment to meet your specific needs.

After you've successfully installed MATLAB on whatever computer platform you're using, you start with the basics and work your way up. By the time you finish working through the examples in this book, you'll be able to perform a range of simple tasks in MATLAB that includes writing scripts, writing functions, creating plots, and performing advanced equation solving. No, you won't be an expert, but you will be able to use MATLAB to meet specific needs in the job environment.

To make absorbing the concepts even easier, this book uses the following conventions:

- ✔ Text that you're meant to type just as it appears in the book is **bold**. The exception is when you're working through a step list: Because each step is bold, the text to type is not bold.

- ✔ When you see words in *italics* as part of a typing sequence, you need to replace that value with something that works for you. For example, if you see "Type *Your Name* and press Enter," you need to replace *Your Name* with your actual name.

- ✔ Web addresses and programming code appear in `monofont`. If you're reading a digital version of this book on a device connected to the Internet, note that you can click the web address to visit that website, like this: `http://www.dummies.com`.

- ✔ When you need to type command sequences, you see them separated by a special arrow like this: File⇨New File. In this case, you go to the File menu first and then select the New File entry on that menu. The result is that you see a new file created.

Foolish Assumptions

You might find it difficult to believe that we've assumed anything about you — after all, we haven't even met you yet! Although most assumptions are indeed foolish, we made these assumptions to provide a starting point for the book.

It's important that you're familiar with the platform you want to use because the book doesn't provide any guidance in this regard. (Chapter 2 does provide MATLAB installation instructions.) To provide you with maximum information about MATLAB, this book doesn't discuss any platform-specific issues. You really do need to know how to install applications, use applications, and generally work with your chosen platform before you begin working with this book.

This book isn't a math primer. Yes, you see lots of examples of complex math, but the emphasis is on helping you use MATLAB to perform math tasks rather than learn math theory. Chapter 1 provides you with a better understanding of precisely what you need to know from a math perspective in order to use this book successfully.

This book also assumes that you can access items on the Internet. Sprinkled throughout are numerous references to online material that will enhance your learning experience. However, these added sources are useful only if you actually find and use them.

Icons Used in This Book

As you read this book, you see icons in the margins that indicate material of interest (or not, as the case may be). This section briefly describes each icon in this book.

Tips are nice because they help you save time or perform some task without a lot of extra work. The tips in this book are timesaving techniques or pointers to resources that you should try in order to get the maximum benefit from MATLAB.

We don't want to sound like angry parents or some kind of maniac, but you should avoid doing anything that's marked with a Warning icon. Otherwise, you might find that your application fails to work as expected, you get incorrect answers from seemingly bulletproof equations, or (in the worst-case scenario) you lose data.

Whenever you see this icon, think advanced tip or technique. You might find these tidbits of useful information just too boring for words, or they could contain the solution you need to get a program running. Skip these bits of information whenever you like.

If you don't get anything else out of a particular chapter or section, remember the material marked by this icon. This text usually contains an essential process or a bit of information that you must know to work with MATLAB successfully.

Beyond the Book

This book isn't the end of your MATLAB experience — it's really just the beginning. We provide online content to make this book more flexible and better able to meet your needs. That way, as we receive email from you, we can address questions and tell you how updates to either MATLAB or its associated add-ons affect book content. In fact, you gain access to all these cool additions:

- ✔ **Cheat sheet:** You remember using crib notes in school to make a better mark on a test, don't you? You do? Well, a cheat sheet is sort of like that. It provides you with some special notes about tasks that you can do with MATLAB that not every other person knows. You can find the cheat sheet for this book at http://www.dummies.com/cheatsheet/matlab. It contains really neat information such as the keyboard shortcuts that you use most often to speed MATLAB use.

✔ **Dummies.com online articles:** A lot of readers were skipping past the parts pages in *For Dummies* books, so the publisher decided to remedy that. You now have a really good reason to read the parts pages — online content. Every parts page has an article associated with it that provides additional interesting information that wouldn't fit in the book. You can find the articles for this book at `http://www.dummies.com/extras/matlab`.

✔ **Updates:** Sometimes changes happen. For example, we might not have seen an upcoming change when we looked into our crystal balls during the writing of this book. In the past, this possibility simply meant that the book became outdated and less useful, but you can now find updates to the book at `http://www.dummies.com/extras/matlab`.

In addition to these updates, check out the blog posts with answers to reader questions and demonstrations of useful book-related techniques at `http://blog.johnmuellerbooks.com/`.

✔ **Companion files:** Hey! Who really wants to type all the code in the book and reconstruct all those plots by hand? Most readers would prefer to spend their time actually working with MATLAB and seeing the interesting things it can do, rather than typing. Fortunately for you, the examples used in the book are available for download, so all you need to do is read the book to learn MATLAB usage techniques. You can find these files at `http://www.dummies.com/extras/matlab`.

Where to Go from Here

It's time to start your MATLAB adventure! If you're completely new to MATLAB, you should start with Chapter 1 and progress through the book at a pace that allows you to absorb as much of the material as possible.

If you're a novice who's in an absolute rush to get going with MATLAB as quickly as possible, you could skip to Chapter 2 with the understanding that you may find some topics a bit confusing later. Skipping to Chapter 3 is possible if you already have MATLAB installed, but be sure to at least skim Chapter 2 so that you know what assumptions we made writing this book.

Readers who have some exposure to MATLAB can save reading time by moving directly to Chapter 5. You can always go back to earlier chapters as necessary when you have questions. However, it's important that you understand how each technique works before moving to the next one. Every technique, coding example, and procedure has important lessons for you, and you could miss vital content if you start skipping too much information.

Part I

Getting Started With MATLAB

In this part . . .

- ✔ Discover why you want to start using MATLAB to speed your calculation.

- ✔ Install MATLAB on your particular system.

- ✔ Start working with MATLAB to become better acquainted with the program.

- ✔ Perform some simple tasks to understand the interface.

- ✔ Become familiar with the MATLAB file system.

Chapter 1

Introducing MATLAB and Its Many Uses

In This Chapter

▶ Understanding how MATLAB fits in as a tool for performing math tasks

▶ Seeing where MATLAB is used today

▶ Discovering how to get the most from MATLAB

▶ Overcoming the MATLAB learning curve

Math is the basis of all our science and even some of our art. In fact, math itself can be an art form — consider the beauty of fractals (a visual presentation of a specialized equation). However, math is also abstract and can be quite difficult and complex to work with. MATLAB makes performing math-related tasks easier. You use MATLAB to perform math-related tasks such as

✔ Numerical computation

✔ Visualization

✔ Programming

This chapter introduces you to MATLAB, an application that performs a variety of math tasks. It helps you understand the role that MATLAB can play in reducing the overall complexity of math and in explaining math-related information to others more easily. You also discover that MATLAB is already used by a number of different organizations to perform real-world tasks in a manner that improves accuracy, efficiency, and consistency. Of course, knowing how you can translate these benefits of MATLAB to your own workplace is important.

Because MATLAB can do so much, it does have a learning curve. This chapter also discusses what you can do to reduce the learning curve so that you become productive much faster. The less time you spend learning about MATLAB, the more time you spend applying math to your particular specialty, and the better the results you achieve. Getting things done quickly and accurately is the overall goal of MATLAB.

Putting MATLAB in Its Place

MATLAB is all about math. Yes, it's a powerful tool and yes, it includes its own language to make the execution of math-related tasks faster, easier, and more consistent. However, when you get right down to it, the focus of MATLAB is the math. For example, you could type 2 + 2 as an equation and MATLAB would dutifully report the sum of 4 as output. Of course, no one would buy an application to compute 2 + 2 — you could easily do that with a calculator. So you need to understand just what MATLAB can do. The following sections help you put MATLAB into perspective so that you better understand how you can use it to perform useful work.

Understanding how MATLAB relates to a Turing machine

Today's computers are mostly Turing machines, named after the British mathematician Alan Turing (1912–1954). The main emphasis of a Turing machine is performing tasks step by step. A single processor performs one step at a time. It may work on multiple tasks, but only a single step of a specific task is performed at any given time. Knowing about the Turing machine orientation of computers is important because MATLAB follows precisely the same strategy. It, too, performs tasks one step at a time in a procedural fashion. In fact, you can download an application that simulates a Turing machine using MATLAB at `http://www.mathworks.com/MATLABcentral/fileexchange/23006-turing-machine-emulator/content/@turing/turing.m`. The code is surprisingly short.

Don't confuse the underlying computer with the programming languages used to create applications for it. Even though the programs that drive the computer may be designed to give the illusion of some other technique, when you look at how the computer works, you see that it goes step by step. If you've never learned how computers run programs, this information is meaningful background. Refer to the nearby sidebar "Understanding how computers work" for a discussion of this important background information.

Understanding how computers work

Many older programmers are geeks who punched cards before TVs had transistors. One advantage of punching cards is getting to physically touch and feel the computer's instructions and data. This physicality gave programmers a good understanding of what happens when a program runs.

Today, the instructions and data are stored as charges of electrons in tiny pieces of silicon too small to be seen through even the most powerful optical microscope. Today's computers can handle much more information much more quickly than early machines. But the way they use that information is basically the same as early computers.

In those old card decks, programmers wrote one instruction on each card. After all the instructions, they put the data cards into a card reader. The computer read a card and the computer did what the card told it to do: Get some data, get more data, add it together, divide, and so on until all the instructions were executed.

A series of instructions is a program. The following figure shows a basic schematic block diagram of how a computer works.

Unchanged from the old days, when cards were read one at a time, computer instructions continue to be read one at a time. The instruction is executed, and then the computer goes to the next instruction. MATLAB executes programs in this manner as well.

It's important to realize that the *flow* of a program can change. Computers can make decisions based on specific criterion, such as whether something is true or false, and take the route indicated for that decision. For example, when the computer has read all the data for a task, the program tells the computer to quit reading data and start doing calculations. Mapping how the computer executes programs is called a *flow chart,* which is similar to a road map with intersections where decisions must be made. MATLAB relies on well-designed flow charts to make it easy to see what the computer will do, when it will do it, and how it will accomplish the required tasks.

The whole concept of a program may seem foreign to many — something that only geeks would ever love — but you've already used the concept of a program before. When using a calculator, you first think of the steps and numbers you want to enter and in what sequence to enter them to solve your problem. A program, including a MATLAB program, is simply a sequence of similar steps stored in a file that the computer reads and executes one at a time. You don't need to fear computer programming — you've probably done something very similar quite often and can do it easily again.

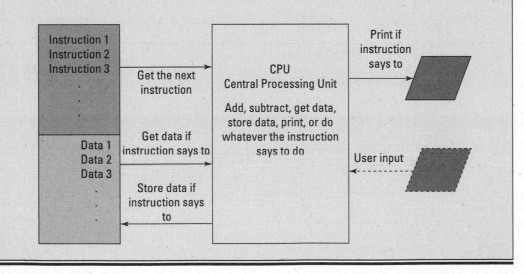

Using MATLAB as more than a calculator

MATLAB is a computer programming language, not merely a calculator. However, you can use it like a calculator, and doing so is a good technique to try ideas that you might use in your program. When you get past the experimentation stage, though, you usually rely on MATLAB to create a program that helps you perform tasks

✔ Consistently

✔ Easily

✔ Quickly

With these three characteristics in mind, the following sections explore the idea of MATLAB's being more than a simple calculator in greater detail. These sections don't tell you everything MATLAB can do, but they do provide you with ideas that you can pursue and use to your own advantage.

Exploring Science, Technology, Engineering, and Mathematics (STEM)

Schools currently have a strong emphasis on Science, Technology, Engineering, and Math (STEM) topics because the world doesn't have enough people who understand these disciplines to get the required work done. Innovation of any sort requires these disciplines, as do many practical trades. MATLAB has a rich and large toolbox for STEM that includes

✔ Statistics

✔ Simulation

✔ Image processing

✔ Symbolic processing

✔ Numerical analysis

Performing simple tasks

Many developers start learning their trade using an older language named Basic. Originally, it was spelled BASIC, for Beginner's All-Purpose Symbolic Instruction Code. The intent behind Basic was to make the language simple. MATLAB retains the simplicity of Basic, but with a much larger toolbox to solve STEM problems. The idea is that you have better things to do than learn how to program using a complex language designed to meet needs that your programs will never address.

Everything has trade-offs. MATLAB is specifically designed to meet the needs of people who use math for learning or to make a living. It gets rid of the complexity found in many other languages and keeps things simple so that you

can focus on your work rather than on the tool you're using to do it. However, in pursuing simplicity, MATLAB is also less flexible than other programming languages, provides fewer advanced features for tasks you'll never perform anyway, and offers fewer generic tools. MATLAB is designed to meet specific needs rather than work as a general-purpose language.

Determining why you need MATLAB

It's important to know *how* to use any application you adopt, but it's equally important to know *when* to use it and what it can actually *do* for your organization. If you don't have a strong reason to use an application, the purchase will eventually sit on the shelf collecting dust. This bit of dust collecting happens far too often in corporations around the world today because people don't have a clear idea of why they even need a particular application. Given that MATLAB can perform so many tasks, you don't want it to just sit on the shelf. The following sections can help you build a case for buying and then using MATLAB in your organization.

Relying on structure for better organization

Writing programs is all about telling the computer to perform a task one step at a time. The better your language tells the computer what to do, the easier the computer will be to use and the less time you'll spend getting it to perform a given task.

Starting with the C and Pascal computer languages, developers began creating structured environments. In such an environment, a map of instructions and decisions doesn't look like a bowl of spaghetti — hard to follow and make sense of — but looks more like a tree, with a trunk and branches that are much easier to follow and understand. MATLAB places a strong emphasis on structure (for example, in the way it organizes data and in the manner in which you write code), which means that you spend a lot more time doing something fun and a lot less time writing code (because the structure means that you work with data in a consistent manner).

Structure does come with a price (there really are trade-offs to everything). Early developers could write an application quickly because they had few rules to follow. Because newer languages do enforce structure (making the code easier to read and update later), you have to spend time learning the rules. The rules are what produce the learning curve in MATLAB that you need to consider as part of working with the product. Make sure that you set realistic goals and establish a timetable that reflects the need to learn programming rules. You can't rush through the MATLAB learning process and expect to do anything useful at the end.

Avoiding the complexity of Object-Oriented Programming (OOP)

You may have heard of Object-Oriented Programming (OOP). It's a discipline that helps developers create applications based on real-world models. Every element of an application becomes an object that has specific characteristics and can perform specific tasks. This technology is quite useful to developers because it helps them create extremely complex applications with fewer errors and less coding time.

However, OOP isn't something you need to know in order to work through various types of math problems. Even though you can solve difficult math problems using languages that do support OOP, STEM users can exploit most of MATLAB's power without OOP. The lack of an OOP requirement means that you can get up and running with MATLAB far faster than you could with a conventional modern programming language and without a loss of the functionality that you need to perform math tasks.

OOP does serve a useful purpose — just not a purpose that you need when creating math models. Leave the complex OOP languages to developers who are creating the software used to access huge databases, or developing a new operating system. MATLAB is designed to make things easy for you.

Using the powerful toolbox

MATLAB provides a toolbox designed to meet the specific needs of STEM users. In contrast to a general programming language, this toolbox provides specific functionality needed to meet certain STEM objectives. Here is just a small sample of the areas that are addressed by the tools you find in the MATLAB toolbox:

- Algebra
- Linear algebra — many equations dealing with many unknowns
- Calculus
- Differential equations
- Statistics
- Curve fitting
- Graphing
- Preparing reports

Reducing programming effort with the fourth-generation language

Programming languages are often rated by their generation. For example, a first-generation language works side by side with the hardware. It's the sort of language that programmers used when computers first appeared on the scene. Nothing is wrong with working directly with the hardware, but you need specialized knowledge to do it, and writing such code is time consuming. A first-generation language is so hard to use that even the developers decided to create something better — second-generation languages! (Second-generation languages, such as Macro Assembler [MASM] are somewhat human-readable, must be assembled into executable code before use, and are still specific to a particular processor.)

Most developers today use a combination of third-generation languages such as C, C++, and Java, and fourth-generation languages such as Structured Query Language (SQL). A third-generation language gives the developer the kind of precise control needed to write exceptionally fast applications that can perform a wide array of tasks. Fourth-generation languages make asking for information easier. For the MATLAB user, the promise of fourth-generation languages means being able to work with collections of data, rather than individual bits and bytes, making it easier for you to focus on the task, instead of the language.

As languages progress from first generation to fourth generation (and beyond), they become more like human language. For example, you might write something like FIND ALL RECORDS WHERE LAST_NAME EQUALS 'SMITH'. It's not quite human language, but close enough that most people can follow it. You tell the computer what to do, but the computer actually decides how to do it. Such languages are useful because they take the burden of interacting with the computer hardware off the language user and place it on the automation that supports the language.

MATLAB employs a fourth-generation language to make your job a lot easier. The language isn't quite human, but it's also a long way away from the machine code that developers used to write to make computers work. Using MATLAB makes you more efficient because the language is specifically designed to meet the needs of STEM users (just as SQL is designed to meet the needs of database administrators and developers who need to access large databases).

Discovering Who Uses MATLAB for Real-World Tasks

An application isn't very useful if you can't perform real-world tasks with it. Many applications out there are curiosities — they may do something interesting, but they aren't really practical. MATLAB is popular among STEM users whose main goal is productively solving problems in their particular field — not problems unique to computer programming. You can find MATLAB used in these kinds of professions:

- ✔ Scientists
- ✔ Engineers
- ✔ Mathematicians
- ✔ Students
- ✔ Teachers
- ✔ Professors
- ✔ Statisticians
- ✔ Control technology
- ✔ Image-processing researchers
- ✔ Simulation users

Of course, most people want to hear about actual users who are employing the product to do something useful. You can find such a list at http://www. mathworks.com/company/user_stories/product.html. Just click the MATLAB entry to see a list of companies that use MATLAB to perform real-world tasks. For example, this list tells you that the Centers for Disease Control (CDC) uses MATLAB for polio virus sequencing (see http://www.mathworks. com/company/user_stories/Centers-for-Disease-Control-and-Prevention-Automates-Poliovirus-Sequencing-and-Tracking. html). You also find that the National Aeronautic and Space Administration (NASA) used MATLAB when creating the model for the X-43 — the craft that recently achieved mach 10 (read about it at http://www.mathworks.com/ company/user_stories/NASAs-X-43A-Scramjet-Achieves-Record-Breaking-Mach-10-Speed-Using-Model-Based-Design.html). The list of companies goes on and on. Yes, MATLAB really is used for important tasks by a large number of companies.

Knowing How to Get the Most from MATLAB

At this point, you may have decided that you absolutely can't get by without obtaining your own personal copy of MATLAB. If that's the case, you really do need to know a little more about it in order to get the most value for your money. The following sections provide a brief overview of the skills that are helpful when working with MATLAB. You don't need these skills to perform every task, but they all come in handy for reducing the overall learning curve and making your MATLAB usage experience nicer.

Getting the basic computer skills

Most complex applications require that you have basic computer skills, such as knowing how to use your mouse, work with menu systems, understand what a dialog box is all about, and perform some basic configuration tasks. MATLAB works like other computer programs you own. It has an intuitive and conventional Graphical User Interface (GUI) that makes using MATLAB a lot easier than employing pad and pen. If you've learned to use a GUI operating system such as Windows or the Mac OS X, and you also know how to use an application such as Word or Excel, you'll be fine.

This book points out MATLAB peculiarities. In addition, you have access to procedures that you can use to make your tasks easier to perform. The combination of these materials will make it easier for you to work with MATLAB even if your computer skills aren't as finely honed as they could be. The important thing to remember is that you can't break anything when working with MATLAB. In fact, we encourage trial and error because it's a great learning tool. If you find that an example doesn't quite work as anticipated, close MATLAB, reopen it, and start the example over again. MATLAB and your computer are both more forgiving than others may have led you to believe.

Defining the math requirements

You need to have the right level of knowledge to use MATLAB. Just as using SQL is nearly impossible without a knowledge of database management, using MATLAB is hard without the proper math knowledge. MATLAB's benefits become evident when applied to trigonometry, exponentials, logarithms, and higher math.

This book assumes that you have a certain level of math knowledge. The math behind the exercises isn't explained to any large degree unless the explanation helps you understand the MATLAB programming language better. However, many sites online cater to math knowledge. For example, you can find a host of tutorials at `http://www.analyzemath.com/`. These tutorials come complete with exercises that help you understand the math behind the MATLAB examples in this book.

Applying what you know about other procedural languages

One of the more significant problems in understanding how to use any language is the procedure. The point was driven home to one fellow at an early age when his teacher assigned his class the task of writing a procedure for making toast. Every student carefully developed a procedure for making toast, and on the day the papers were turned in, the teacher turned up with a loaf of bread and a toaster. She dutifully followed the instructions each child provided to the letter. All the children failed at the same point. Yes, they forgot to take the bread out of the wrapper. You can imagine what it was like trying to shove a single piece of bread into the toaster when the piece was still in the wrapper along with the rest of the bread.

Programming can be (at times) just like the experiment with the toast. The computer takes you at your word and follows to the letter the instructions you provide. The results may be not what you expected, but the computer always follows the same logical course. Having previous knowledge of a procedural language, such as C, Java, C++, or Python, will help you understand how to write MATLAB procedures as well. You have already developed the skill required to break instructions into small pieces and know what to do when a particular piece is missing. Yes, you can use this book without any prior programming experience, but the prior experience will most definitely help you get through the chapters must faster and with fewer errors.

Understanding how this book will help you

This is a *For Dummies* book, so it takes you by the hand to explore MATLAB and make it as easy to understand as possible. The goal of this book is to help you use MATLAB to perform at least simple feats of mathematical magic. It won't make you a mathematician and it won't help you become a developer — those are topics for other books. When you finish this book, you will know how to use MATLAB to explore STEM-related topics.

Make sure that you also check out the blog for this book at http://blog.johnmuellerbooks.com/categories/263/matlab-for-dummies.aspx. This is the place to look for updates and additional information. Also, the blog offers answers to commonly asked questions — questions asked by people just like you. You can also write the authors: John (John@JohnMuellerBooks.com) and Jim (jsiz@tjc.edu) with your book-related questions. We want to ensure that you have a great reading experience and can get everything possible from this book.

Getting Over the Learning Curve

Even easy programming languages have a learning curve. If nothing else, you need to discover the techniques that developers use to break tasks into small pieces, ensure that all the pieces are actually there, and then place the pieces in a logical order. Creating an orderly flow of steps that the computer can follow can be difficult, but this book leads you through the process a step at a time.

To help you understand MATLAB, this book compares how to accomplish a task in MATLAB with something you're used to, such as a spreadsheet or calculator. You learn by doing. Try the examples in this book and invent some of your own. Try variations and experiment. MATLAB's not too tough — you, too, can discover how to use MATLAB.

Chapter 2

Starting Your Copy of MATLAB

In This Chapter

▶ Obtaining and installing your copy of MATLAB

▶ Starting MATLAB and working with the interface

*B*efore you can use MATLAB to do anything productive, you need a copy of it installed on your system. Fortunately, you can obtain a free trial version that lasts 30 days. If you're diligent, you can easily complete this book in that time and know for certain whether you want to continue using MATLAB as a productivity aid. The point is that you need a good installation, and this book helps you obtain that goal.

 After you have MATLAB installed, it's important to introduce yourself to the interface. This chapter provides you with an overview of the interface, not a detailed look at every feature. However, overviews are really important because working with lower-level interface elements is hard if you don't have the big picture. You may actually want to mark this chapter in some way so that you can refer back to the interface information.

Installing MATLAB

A problem that anyone can encounter is getting a bad product installation or simply not having the right software installed. When you can't use your software properly, the entire application experience is less than it should be. The following sections guide you through the MATLAB installation so that you can have a great experience using it.

Discovering which platforms MATLAB supports

Before you go any further, you need to verify that your system will actually run MATLAB. At a minimum, you need 3GB of free hard drive space and 2GB of RAM to use MATLAB effectively. (It can run on systems with fewer resources, but you won't be happy with the performance.) You also need to know which platforms MATLAB supports. You can use it on these systems:

- ✔ Windows (3GB free disk space, 2GB RAM)
  - Windows 8.1
  - Windows 8
  - Windows 7 Service Pack 1
  - Windows Vista Service Pack 2
  - Windows XP Service Pack 3
  - Windows XP x64 Edition Service Pack 2
  - Windows Server 2012
  - Windows Server 2008 R2 Service Pack 1
  - Windows Server 2008 Service Pack 2
  - Windows Server 2003 R2 Service Pack 2
- ✔ Mac OS X
  - Mac OS X 10.9 (Mavericks)
  - Mac OS X 10.8 (Mountain Lion)
  - Mac OS X 10.7.4+ (Lion)
- ✔ Linux
  - Ubuntu 12.04 LTS, 13.04, and 13.10
  - Red Hat Enterprise Linux 6.*x*
  - SUSE Linux Enterprise Desktop 11 SP3
  - Debian 6.*x*

Linux users may find that other distributions work. However, the list of Linux systems represents those that are tested to work with MATLAB. If you try MATLAB on your unlisted Linux system and find that it works well, please let John know (at John@JohnMuellerBooks.com) and he'll mention these other systems in a blog post. The point is that you really do need to have the right platform to get good results with MATLAB. You can always obtain the current minimum requirements for MATLAB at http://www.mathworks.com/support/sysreq/current_release/index.html.

Getting your copy of MATLAB

Before you can work with MATLAB, you need a copy installed on your system. Fortunately, you have a number of methods at your disposal. Here are the three most common ways of getting MATLAB:

✔ Get the trial version from `https://www.mathworks.com/programs/trials/trial_request.html`.

✔ Obtain a student version of the product from `https://www.mathworks.com/academia/student_version/`.

✔ Buy a copy from `http://www.mathworks.com/pricing-licensing/index.html`.

In most cases, you need to download the copy of MATLAB or the MATLAB installer onto your system after you fill out the required information to get it. Some users choose to receive a DVD in the mail instead of downloading the product online. No matter which technique you use, you eventually get a copy of MATLAB to install.

Performing the installation

The method you use to install MATLAB depends on the version you obtain and the media used to send it to you. For example, there is a method for installing MATLAB from DVD and an entirely different method when you want to download the installer and use an Internet connection. Administrators and users also have different installation procedures. Use the table at `http://www.mathworks.com/help/install/ug/choose-installation-procedure.html` to determine which installation procedure to use.

MathWorks provides you with substantial help in performing the installation. Before you contact anyone, be sure to look through the materials on the main installation page at `http://www.mathworks.com/help/install/index.html`. It's also possible to obtain installation help at `http://www.mathworks.com/support/install-matlab.html`. Take the time to review the material that MathWorks provides before you push the panic button. Doing so will save time and effort.

Activating the product

After you complete the MATLAB installation, you must activate the product. Activation is a verification process. It simply means that MathWorks verifies that you have a valid copy of MATLAB on your system. With a valid copy, you obtain support such as updates to your copy of MATLAB as needed.

As with installation, you have a number of activation types to use with MATLAB that depend on the product version and how you're using the product. The chart at `http://www.mathworks.com/help/install/license/activation-types.html` tells you about these activation types and helps you make a selection. The matrix at `http://www.mathworks.com/help/install/license/`

`license-option-and-activation-type-matrix.html` tells you whether your particular version of MATLAB supports a specific activation type. For example, the individual license doesn't support the Network Named User activation type.

MATLAB automatically asks you about activation after the installation process is complete. You don't need to do anything special. However, you do want to consider the type of activation you want to perform — which type of activation will best meet your needs and those of your organization.

Meeting the MATLAB Interface

Most applications have similar interface functionality. For example, if you click a button, you expect something to happen. The button usually contains text that tells you what will happen when you click it, such as closing a dialog box by clicking OK or Cancel. However, the similarities aren't usually enough to tell you everything you need to know about the interface. The following sections provide an overview of the MATLAB interface so that you can work through the chapters that follow with greater ease. These sections don't tell you everything about the interface, but you do get enough information to feel comfortable using MATLAB.

Starting MATLAB for the first time

When you start MATLAB for the first time (after you activate it), you see a display containing a series of blank windows. It's not all that interesting just yet because you haven't done anything with MATLAB. However, each of the windows has a special purpose, so it's important to know which window to use when you want to perform a task.

It's possible to arrange the windows in any order needed. Figure 2-1 shows the window arrangement used throughout the book, which may not precisely match your display. The "Changing the MATLAB layout" section of this chapter tells you how to rearrange the windows so that you can see them the way that works best when you work. Here is a brief summary of the window functionality.

✔ **Home tab:** The Home tab of the *Ribbon interface* (a bar that provides access to various MATLAB features, such as creating a new script or importing data) is where you find most of the icons you use to create and use MATLAB formulas. It's the tab you use most often. Also on the interface is a Plots tab (for creating graphic presentations of your formulas) and an Apps tab (for obtaining new applications to use with MATLAB). MATLAB calls the Ribbon interface the *Toolstrip,* so that's the name we use throughout the book.

Current Folder Window   Minimize Toolstrip

Home Tab   Address Field   Quick Access Toolbar   Workspace window

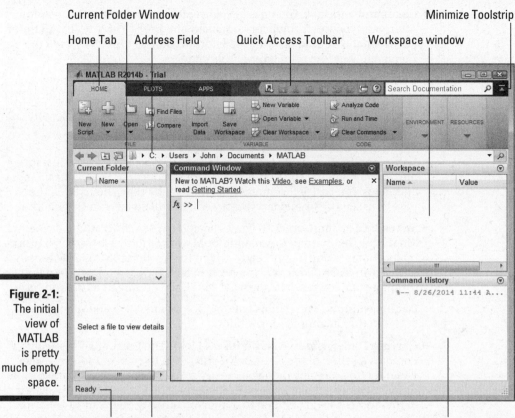

Figure 2-1:
The initial
view of
MATLAB
is pretty
much empty
space.

Status Bar   Details Window   Command Window   Command History Window

✔ **Quick Access toolbar:** The Quick Access toolbar (QAT) provides access
to the MATLAB features that you use most often. Finding icons on the
QAT is often faster and easier than looking them up on the Toolstrip.

You can change the QAT to meet your needs. To add an icon to the QAT,
right-click its entry in the Toolstrip and choose Add to Quick Access
Toolbar from the context menu. If you want to remove an icon from the
QAT, right-click its entry in the QAT and choose Remove from the Quick
Access Toolbar from the context menu.

✔ **Minimize Toolstrip:** If you find that the Toolstrip is taking up too much
space, you can click the Minimize Toolstrip icon to remove it from view.
To restore the Toolstrip, simply click the Minimize Toolstrip icon again.

When the Toolstrip is minimized, you can still see the three tabs —
Home, Plots, and Apps. Click a tab to reveal the Toolstrip long enough
to use a MATLAB feature. As soon as you select a Toolstrip feature or
click in another MATLAB area, the Toolstrip disappears again. Using this
technique allows you full access to the MATLAB features but keeps the
Toolstrip hidden to save space.

- ✔ **Command window:** You type formulas and commands in this window. After you type the formula or command and press Enter, MATLAB determines what it should do with the information you typed. You see the Command window used later in this chapter.

- ✔ **Workspace window:** The Workspace window contains the results of any tasks you ask MATLAB to perform. It provides a scratchpad of sorts that you use for calculations. The Workspace window and Command window work hand in hand to provide you with a complete view of the work you perform using MATLAB.

- ✔ **Command History window:** In some cases, you want to reissue a formula or command. The Command History window acts as your memory and helps you restore formulas and commands that you used in the past. You see the Command History window used later in this chapter.

- ✔ **Status bar:** It's important to know the current MATLAB state — whether MATLAB is ready to perform additional work or not. The status bar normally contains one word, Ready, which tells you that MATLAB is ready to perform tasks. However, you need to watch this window when performing complex tasks to see what MATLAB is doing at any given time.

- ✔ **Details window:** The Details window shows specifics about any file you select in the Current Folder window.

- ✔ **Current Folder window and Address Field:** The Current Folder window contains a listing of the files you've created in the current folder — files you'd use to store any data you create in MATLAB, along with any scripts or functions you'd use to manipulate data). The Current Folder is listed in the Address Field text box that appears directly below the Toolstrip. Changing the Address Field text box content also changes the content of the Current Folder window.

Employing the Command window

The Command window is where you perform most of your experimentation. This chapter shows how to perform really simple tasks using the Command window, but as the book progresses, you see that the Command window can do quite a lot for you. The following sections describe some of the ways in which you can use the Command window to learn more about MATLAB.

Typing a really simple command

You can type any formula or command desired in the Command window and see a result. Of course, it pays to start with something really simple so that you can get the feel of how this window works. Type 2 + 2 and press Enter in the Command window. You see the results shown in Figure 2-2.

Notice that it isn't just the Command window that is affected when you type a formula or command. Other windows are changed as well. Here are the all windows affected:

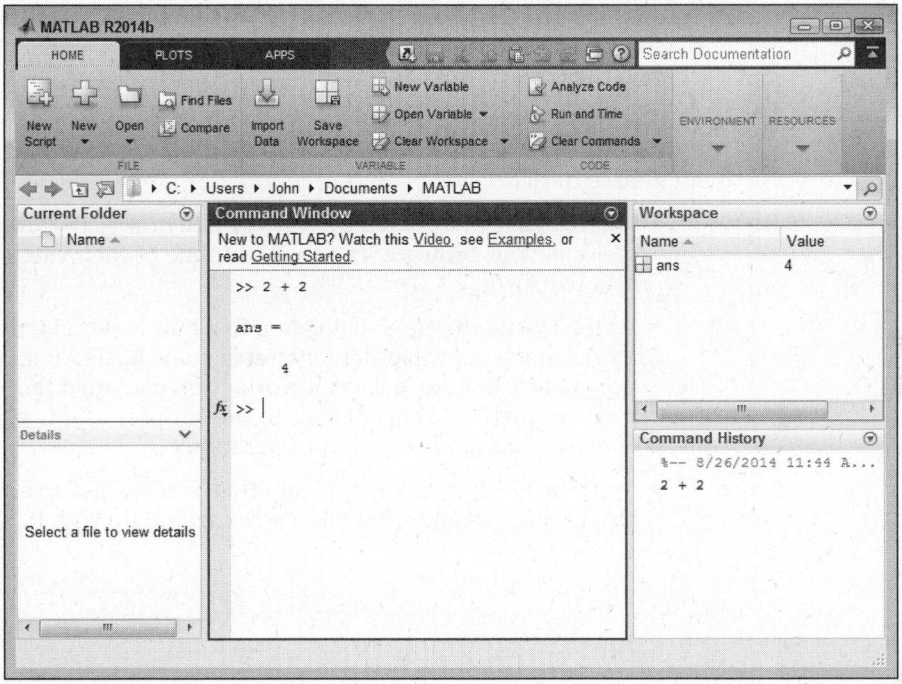

Figure 2-2:
A very simple command in MATLAB.

✔ **Command window:** Receives the output of the formula 2 + 2, which is ans = 4. MATLAB assigns the output of the formula to a variable named ans. *Variables* are boxes (pieces of memory) in which you can place data. In this case, the box contains the number 4.

✔ **Workspace window:** Contains any variables generated as the result of working in the Command window. In this case, the Workspace window contains a variable named ans that holds a value of 4.

Notice that the variable can't contain any other value than 4 because the Min column also contains 4, as does the Max column. When a variable can contain a range of values, the minimum value that it can contain appears in the Min column and the maximum value that it can contain appears in the Max column. The Value column always holds the current value of the variable.

✔ **Command History window:** Displays the series of formulas or commands that you type, along with the date and time you typed them. You can replay a formula or command in this window. Just select the formula or command that you want to use from the list to replay it.

Getting additional help

At the top of the Command window, notice three links that you can use to quickly get additional help in using MATLAB. Each link helps you in a different way, so you can use the links as needed to meet your specific needs. Here is an overview of what each link provides:

✔ **Watch This Video:** Opens a tutorial in your browser. The video provides a brief introduction to MATLAB. Simply watch it for a visual presentation of how to work with MATLAB.

✔ **See Examples:** Displays a Help dialog box that contains an assortment of examples that you can try, as shown in Figure 2-3. The examples take a number of forms:

- **Video:** Displays a guided presentation of how to perform a task that opens in your browser. The length of time of each video is listed next to its title.

- **Script:** Opens the Help dialog box to a new location that contains an example script that demonstrates some MATLAB feature and an explanation of how the script works. You can open the script and try it yourself. Making changes to the script is often helpful to see how the change affects script operation.

- **App:** Starts a fully functional app that you can use to see how MATLAB works and what you can expect to do with it.

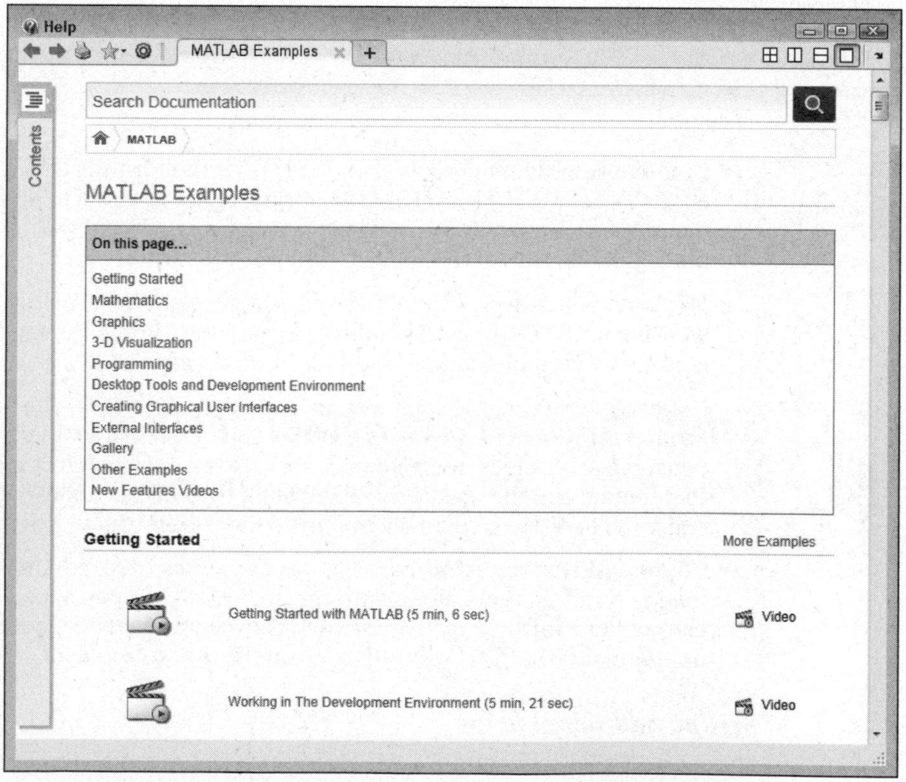

Figure 2-3: The examples give you practical experience using MATLAB.

✔ **Read Getting Started:** Displays a Help dialog box that contains additional information about MATLAB, such as the system requirements, as shown in Figure 2-4. You also gain access to a number of tutorials.

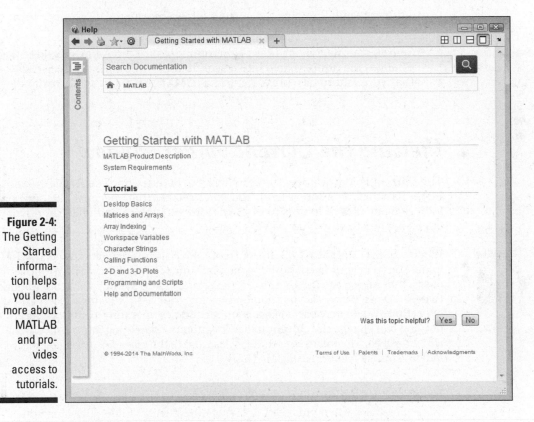

Figure 2-4:
The Getting Started information helps you learn more about MATLAB and provides access to tutorials.

Using the Current Folder toolbar

The Current Folder toolbar helps you navigate the Current Folder window with greater precision. Here is a description of each of the toolbar elements when viewed from left to right on the toolbar:

✔ **Back:** Moves you back one entry in the file history listing. MATLAB retains a history of the places you visit on the hard drive. You can move backward and forward through this list to get from one location to another quite quickly.

✔ **Forward:** Moves you forward one entry in the file history listing.

✔ **Up One Level:** Moves you one level up in the directory hierarchy. For example, when viewing Figure 2-9, if you are currently in the \MATLAB\ Chapter02 folder, clicking this button takes you to the \MATLAB folder.

> ✔ **Browse for Folder:** Displays a Select a New Folder dialog box that you can use to view the hard drive content. Highlight the folder you want to use and click Select to change the Current Folder window location to the selected folder.
>
> ✔ **Address field:** Contains the current folder information. Type a new value and press Enter to change the folder.
>
> ✔ **Search (**the Magnifying Glass icon to the right of the Address field)**:** Changes the Address field into a search field. Type the search criteria that you want to use, press Return, and MATLAB displays the results for you in the Current Folder window.

Viewing the Current Folder window

The Current Folder window (refer to Figure 2-1) really does show the current folder listed in the Address field. You don't see anything because the current folder has no files or folders to display. However, you can add files and folders as needed to store your MATLAB data.

When you first start MATLAB, the current folder always defaults to the MATLAB folder found in your user folder for the platform of your choice. For Windows users, that means the `C:\Users\<User Name>\Documents\MATLAB` folder (where `<User Name>` is your name). Burying your data way down deep in the operating system may seem like a good idea to the operating system vendor, but you can change the current folder location to something more convenient when desired. The following sections describe techniques for managing data and its storage location using MATLAB.

Temporarily changing the current folder

There are times when you need to change the current folder. Perhaps your data is actually stored on a network drive, you want to use a shared location so that others can see your data, or you simply want to use a more convenient location on your local drive. The following steps help you change the current folder:

1. Click Set Path in the Environment group on the Toolstrip's Home tab.

You see the Set Path dialog box shown in Figure 2-5.

This dialog box lists all the places the MATLAB searches for data, with the default location listed first. You can use these techniques to work with existing folders (go to Step 3 when you're finished):

- To set an existing folder as the default folder, highlight the folder in the list and click Move to Top.

- To stop using an existing folder, highlight the folder in the list and click Remove.

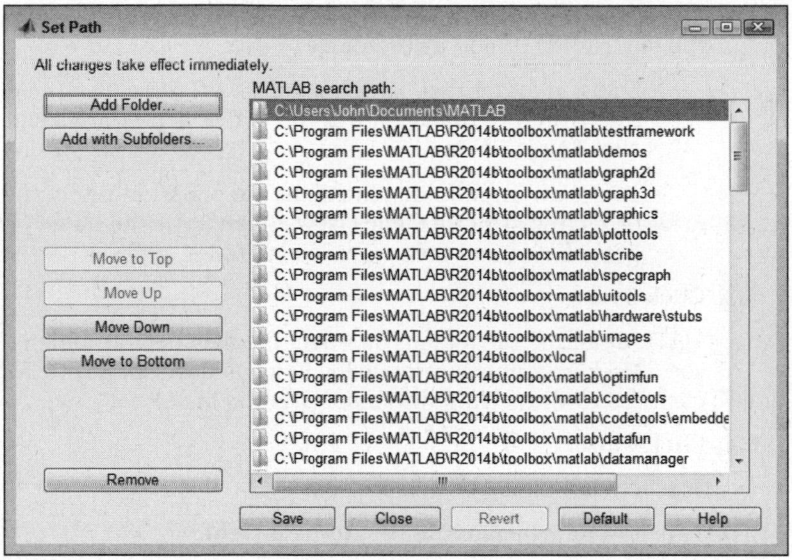

Figure 2-5: The Set Path dialog box contains a listing of folders that MATLAB searches for data.

2. **Click Add Folder.**

   You see the Add Folder to Path dialog box, as shown in Figure 2-6.

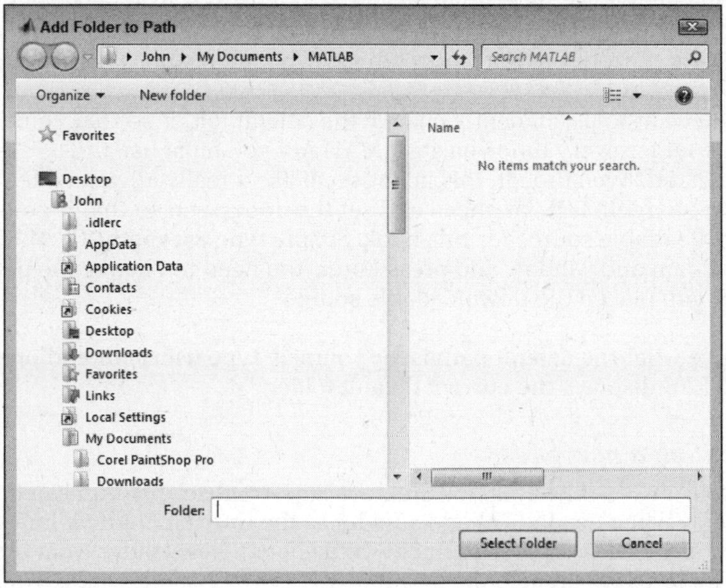

Figure 2-6: Look for the folder you want to use or add a new one.

This dialog box lets you choose an existing folder that doesn't appear in the current list or add a new folder to use:

- To use a folder that exists on your hard drive, use the dialog box's tree structure to navigate to the folder, highlight its entry, and then click Select Folder.

- To create a new folder, highlight the parent folder in the dialog box's tree structure, click New Folder, type the name of the folder, press Enter, and then click Select Folder.

3. **Click Save.**

MATLAB makes the folder you select the new default folder. (You may see a User Account Control dialog box when working with Windows; click Yes to allow Windows to perform the task.)

4. **Click Close.**

The Set Path dialog box closes.

5. **Type the new location in the Address field.**

The Current Folder display changes to show the new location.

Permanently changing the default folder

The default folder is the one that MATLAB uses when it starts. Setting a default folder saves you time because you don't have to remember to change the current folder setting every time you want to work. If you have your default folder set to the location from which you work most of the time, you can usually get right to work and not worry too much about locations on the hard drive.

If you want to permanently change the default folder so that you see the same folder every time you start MATLAB, you must use the userpath() command. Even though this might seem like a really advanced technique, it isn't hard. In fact, go ahead and set the userpath so that it points to the downloadable source for this book. Simply type **userpath('C:\MATLAB')** in the Command window and press Enter. You need to change the path to wherever you placed the downloadable source.

To see what the default path is for yourself, type **userpath** and press Enter. MATLAB displays the current default folder.

Creating a new folder

Organizing the files that you create is important so that you can find them quickly when needed. To add a folder to the Current Folder window, right-click any clear area in the window and choose New Folder from the context menu. MATLAB creates the new folder for you. Type the name you want to use for the new folder and press Enter.

Each chapter in this book uses a separate folder to store any files you create. When you obtain the downloadable source from the publisher's site (http://www.dummies.com/extras/matlab), you find the files for this chapter in the \MATLAB\Chapter02 folder. Every other chapter will follow the same pattern.

Saving a formula or command as a script

After you create a formula or command that you want to use to perform a number of calculations, be sure to save it to disk. Of course, you can save anything that you want to disk, even the simple formula you typed earlier in this chapter. The following steps help you save any formula or command that you want to disk so that you can review it later:

1. **Choose a location to save the formula or command in the Address field.**

2. **Right-click the formula or command that you want to save in the Command History window and choose Create Script from the context menu.**

   You see the Editor window, as shown in Figure 2-7. The script is currently untitled, so you see the script name as Untitled*. (Figure 2-7 shows the Editor window undocked so you can see it with greater ease — the "Changing the MATLAB layout" section of this chapter tells how to undock windows so you can get precisely the same look.)

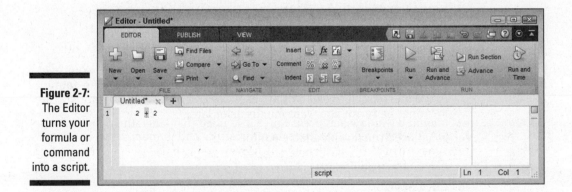

Figure 2-7: The Editor turns your formula or command into a script.

If you want to select multiple commands to place in a script, you can choose them by clicking on the first command, and then using Ctrl+Click to select any additional commands. Each time you Ctrl+Click on a command, MATLAB highlights its entry. The commands will appear in the script file in the same order in which they appear in the Command History window, rather than in the order in which you click on them.

3. **Click Save on the Editor tab.**

   You see the Select File for Save As dialog box, as shown in Figure 2-8.

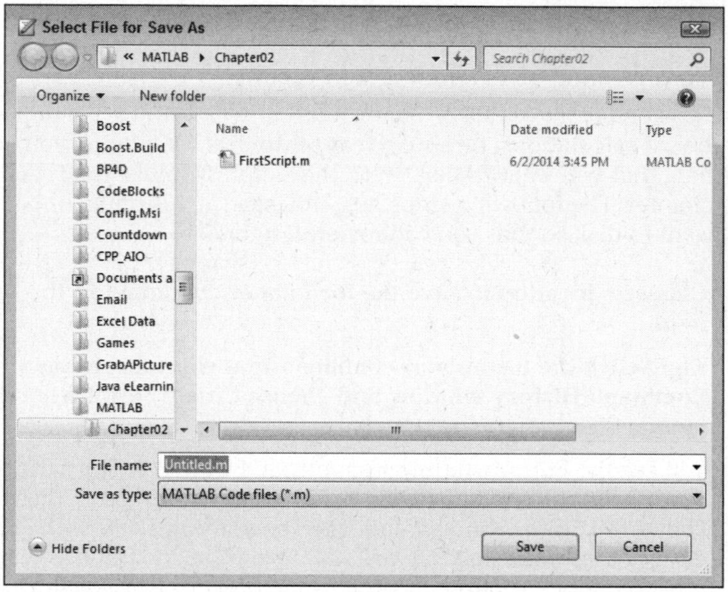

Figure 2-8:
Choose a
location to
save your
script and
provide a
filename
for it.

4. **In the left pane, highlight the location you want to use to save the file.**

5. **Type a name for the script in the File Name field.**

   The example uses `FirstScript.m`. However, when you save your own scripts, you should use a name that will help you remember the content of the file. Descriptive names are easy to remember and make precisely locating the script you want much easier later.

   MATLAB filenames can contain only letters and numbers. You can't use spaces in a MATLAB filename. However, you can use the underscore in place of a space.

6. **Click Save.**

   MATLAB saves the script for you so that you can reuse it later. The title bar changes to show the script name and its location on disk.

7. **Close the Editor window.**

   The Current Folder window displays the folder and script file that you've created using the previous steps in this chapter, as shown in Figure 2-9.

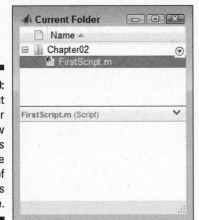

Figure 2-9:
The Current
Folder
window
always
shows the
results of
any changes
you make.

Running a saved script

You can run any script by right-clicking its entry in the Current Folder window and choosing Run from the context menu. When you run a script, you see the script name in the Command window, the output in the Workspace window, and the actual command in the Command History window, as shown in Figure 2-10.

Saving the current workspace to disk

Sometimes you might want to save your workspace to protect work in progress. The work may not be ready to turn into a script, but you want to save it before quitting for the day or simply to ensure that any useful work isn't corrupted by errors you make later.

To save your workspace, click Save Workspace in the Variable group of the Toolstrip's Home tab. You see a Save to MAT-file dialog box that looks similar to the Select File for Save As dialog box (refer to Figure 2-8). Type a filename for your workspace, such as **FirstWorkspace.mat**, and click Save to save it.

Workspaces use a .mat extension, while scripts have a .m extension. Make sure that you don't confuse the two extensions. In addition, workspaces and scripts use different icons so that you can easily tell them apart in the Current Folder window.

Changing the MATLAB layout

The MATLAB layout is designed to make experimentation easy and comfortable for you. However, you may find after a while that it really doesn't meet your needs. Fortunately, you can reconfigure the MATLAB layout to any configuration you want. The following sections provide ideas on how you can reconfigure the MATLAB layout.

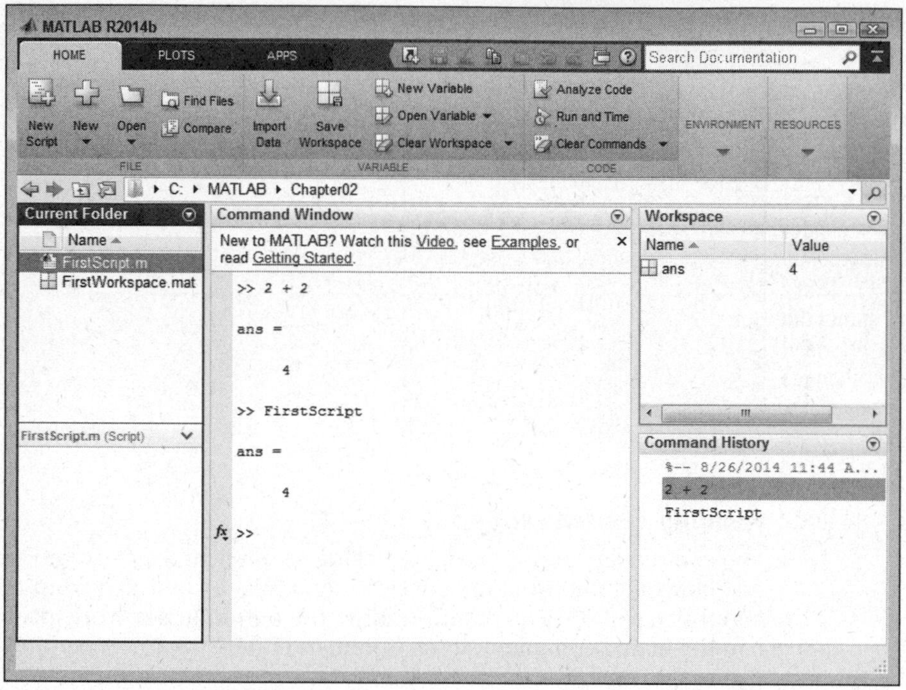

Figure 2-10:
Running a
script shows
its name and
results.

Minimizing and maximizing windows

Sometimes you need to see more or less of a particular window. It's possible to simply resize the windows, but you may want to see more or less of the window than resizing provides. In this case, you can minimize the window to keep it open but completely hidden from view, or maximize the window to allow it to take up the entire client area of the application.

On the right side of the title bar for each window, you see a down arrow. When you click this arrow, you see a menu of options for that window, such as the options shown in Figure 2-11 for the Current Folder window. To minimize a window, choose the Minimize option from this menu. Likewise, to maximize a window, choose the Maximize option from the menu.

Eventually, you want to change the window size back to its original form. The Minimize or Maximize option on the menu is replaced by a Restore option when you change the window's setup. Select this option to restore the window to its original size.

Opening and closing windows

In some cases, you may no longer need the information found in a particular window. When this happens, you can close the window. MATLAB doesn't actually destroy the window contents, but the window itself is no longer accessible. To close a window that you don't need, click the down arrow on the right side of the window and choose Close from the menu.

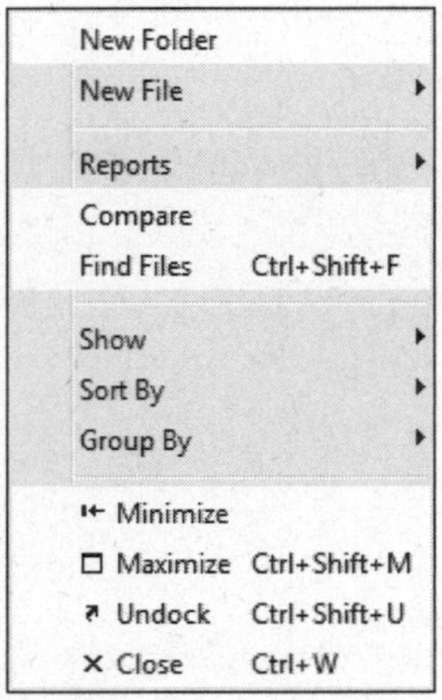

New Folder	
New File	▶
Reports	▶
Compare	
Find Files	Ctrl+Shift+F
Show	▶
Sort By	▶
Group By	▶
⟻ Minimize	
☐ Maximize	Ctrl+Shift+M
⬈ Undock	Ctrl+Shift+U
✕ Close	Ctrl+W

Figure 2-11:
The window menus contain options for changing the appearance of the window.

After you close a window, the down arrow is no longer accessible, so you can't restore a closed window by using the menu options shown in Figure 2-11. To reopen a window, you click the down arrow on the Layout button in the Environment group of the Home tab. You see a list of layout options like the ones shown in Figure 2-12.

The Show group contains a listing of windows. Each window with a check mark next to it is opened for use (closed windows have no check mark). To open a window, click its entry. Clicking the entry places a check next to that window and opens it for you. The window is automatically sized to the size it was the last time you had it open.

You can also close windows using the options on the Layout menu. Simply click the check next to a window entry to close it.

Docking and undocking windows

Many people have multiple monitors attached to their systems. It's often more efficient to perform the main part of your work on your main monitor and move supplementary windows to a second monitor. However, you really can't move a window until you undock the window from MATLAB so that you can move just that window to another location.

To undock a window, click the down arrow on the right side of its title bar and choose Undock from the menu. The window becomes a separate entity, much like the Current Folder window shown back in Figure 2-9. You can move the undocked window anywhere you want it, including to a second monitor.

At some point, you may decide that you want MATLAB to have all its windows in one place again. In this case, you click the down arrow on the right side of the window's title bar and choose Dock from the menu. MATLAB places the window precisely where it was before you undocked it. However, the window may not return to its original size — you may need to resize it to make it fit as it did before.

Choosing an existing layout

One of the potential problems of changing your layout is that it may cause MATLAB to become nearly unusable. Rather than spend a lot of time trying to get the original layout back, you can simply choose an existing layout. To perform this task, click the down arrow on the Layout button in the Environment group of the Home tab and choose one of the Select Layout options. The Default entry returns MATLAB to the same state it was in when you started it the first time.

Saving a new layout

After you find the perfect layout for your needs, you want to save it to disk so that you can easily restore it later should the MATLAB display become disorganized (perhaps you've moved things about to perform a particular task). To perform this task, click the down arrow on the Layout button in the Environment group of the Toolstrip's Home tab and choose Save Layout. You see the Save Layout dialog box. Type the name of the layout in the space provided and click OK. The layout now becomes available in the Select Layout section of the Layout menu.

Chapter 3

Interacting with MATLAB

· ·

In This Chapter

▶ Performing basic calculations

▶ Creating more complex calculations

▶ Interacting with variables

▶ Using MATLAB functions

▶ Overcoming errors

▶ Obtaining additional help

· ·

*Y*ou can interact with MATLAB in a lot of ways and you'll experience quite a few of them as the book progresses. However, it pays to start out slowly to build your skills. This chapter presents an overview of the sorts of things you can do with MATLAB. Use this chapter to get started with a product that can really perform complex tasks with aplomb.

Although you probably won't spend a lot of time using MATLAB as a calculator for even complex calculations, you can do so. Rather than view this kind of use as a waste of time, however, view it as a means of practicing as well as experimentation. Sometimes playing with a product produces unexpected outcomes that can help you in your daily work. To that end, this chapter introduces you to MATLAB through the use of direct calculation entries.

Another type of interaction with MATLAP covered in this chapter occurs through variables. Think of *a variable* as a kind of storage box. You put data into a variable so that you can store that data for a while. Later, when you need the data again, you take it out of the variable, do something with it, and put it back in again. Variables have nothing mystical or difficult about them; in fact, you use variables all the time in real life. For example, you could view your refrigerator as a kind of variable. You put the bag of apples inside to store them for a short time, take the bag out to remove an apple to eat, and put the rest of the bag back into the refrigerator (minus one apple). The point is that developers make a big deal out of fancy terms (that you unfortunately also need to use in order to talk with them), but in reality there isn't anything odd about them. You get a fuller explanation of variables as part of this chapter.

In the process of interacting with MATLAB, you'll make mistakes. Of course, everyone makes mistakes. MATLAB won't blow up if you make a mistake, and your computer won't up and run away. Mistakes are part of the learning process, so you need to embrace them. In fact, most of the greatest people in history made a ton of mistakes (see Defining the Benefits of Failure at `http://blog.johnmuellerbooks.com/2013/04/26/defining-the-benefits-of-failure/`). This book assumes that you're going to make mistakes, so part of this chapter discusses how to recover from them. Knowing how to recover means that you don't have to worry about making a mistake because you can always start fresh.

And finally in this chapter is the topic of additional resources for finding help. No one wants to reinvent the wheel, and a lack of progress can become discouraging after a while. That's why you'll definitely want to know where to find help on using MATLAB. The final section of this chapter discusses techniques you can use to obtain additional help. Working through issues with MATLAB on your own is important because that's how you learn. However, after you've worked through the issues for a while, you also need to know where to get additional help.

Using MATLAB as a Calculator

MATLAB performs math tasks incredibly well. Sometimes people get so caught up in "what else" an application can do that they miss the most interesting facts that are staring them right in the face. The following sections help you understand MATLAB as a calculator so that you can use it for experimentation purposes.

Entering information at the prompt

References to using the prompt appear a few times in previous chapters, but those chapters don't fully explain it. The *prompt* is that place where you type formulas, commands, or functions or perform tasks using MATLAB. It appears in the Command window. Normally, the prompt appears as two greater-than signs (>>). However, when working with some versions of MATLAB, you might see EDU>> (for the student version) or Trial>> (for the trial version) instead. No matter what you see as a prompt, you use it to know where to type the information described in this book.

Chapter 2 shows you how to use something called the userpath() function to alter the permanent path that MATLAB uses when starting up. In this chapter, we introduce you to a useful command known as clc. Try it now:

Type **clc** and press Enter at the MATLAB prompt. If the Command window contains any information, MATLAB clears it for you.

The userpath() function is called a *function* because it uses parentheses to hold the data — also called *arguments* — you send to MATLAB. The clc command is a *command* because you don't use parentheses with it. Whether something is a function or a command depends on how you use it. The usage is called the function or command *syntax* (the grammar used to tell MATLAB what tasks to perform). It's possible to use userpath() in either Function or Command form. To avoid confusion, the book usually relies on function syntax when you need to provide arguments, and command syntax when you don't. So, when you see parentheses, you should also expect to provide input with the *function call* (the act of typing the function and associated arguments, and then pressing Enter).

MATLAB is also *case sensitive.* That sounds dangerous, but all it really means is that CLC is different from Clc, which is also different from clc. Type **CLC** and press Enter at the MATLAB prompt. You see an error message like the one shown in Figure 3-1. (MATLAB will also suggest the correct command, clc, but ignore the advice for right now by highlighting clc and pressing Delete.) Next, type **Clc** and press Enter at the MATLAB prompt. This time, you see the same error because you made the "same" mistake — at least in the eyes of MATLAB. If you see this error message, don't become confused simply because MATLAB didn't provide a clear response to what you typed — just retype the command, being sure to type the command exactly as written.

Figure 3-1:
MATLAB is case sensitive, so CLC, Clc, and clc all mean different things.

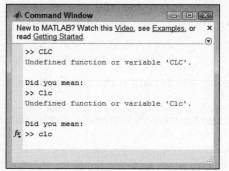

Notice also the "Did you mean:" text that appears after the error message. Normally, MATLAB tries to help you fix any errors. In some cases, MATLAB can't figure out what's wrong, so it won't provide any alternatives for you. (In other cases, MATLAB provides an alternative, so you need to check the prompt to determine whether help is available.) Because MATLAB was able to provide the correct command in this case, simply press Enter to clear the Command window.

Look in the Command History window. Notice that there is a red line next to each of the errant commands you typed. These red lines tell you when you shouldn't use a command or function again because it produced an error the first time. You should also avoid adding errant commands and functions to any scripts you create.

Entering a formula

To enter a formula, you simply type it. For example, if you type **2 + 2** and press Enter, you get an answer of 4. Likewise, if you type **2 * pi * 6378.1** and press Enter, you get the circumference of the earth in km (see http://nssdc.gsfc.nasa.gov/planetary/factsheet/earthfact.html for a list of Earth statistics, including radius). The second formula uses a predefined constant, pi, which equals 3.1416. MATLAB actually defines a number of predefined constants that you can use when entering a formula:

- ✔ **ans:** Contains the most recent temporary answer. MATLAB creates this special temporary variable for you when you don't provide a variable of your own.

- ✔ **eps:** Specifies the accuracy of the floating-point precision (epsilon), which defaults to 2.2204e-16.

- ✔ **i:** Contains an imaginary number, which defaults to 0.0000 + 1.0000i.

- ✔ **Inf:** Defines a value of infinity, which is any number divided by 0, such as 1 / 0.

- ✔ **NaN:** Specifies that the numerical result isn't defined (Not a Number).

- ✔ **pi:** Contains the value of pi, which is 3.1416 when you view it onscreen. Internally, MATLAB stores the value to 15 decimal places so that you're assured of accuracy.

Whenever you type a formula and press Enter, you get an output that specifies the value of ans, which is a temporary value that holds the answer to your question. For example, try typing **2 * pi * 6378.1** and pressing Enter. You see the circumference of the Earth, as shown in Figure 3-2.

Copying and pasting formulas

With MATLAB, you can copy and paste formulas that you create into other documents (such as a script or function file, or to another application). To begin, you highlight the information you want to copy. Use one of these methods to copy the text after you highlight it:

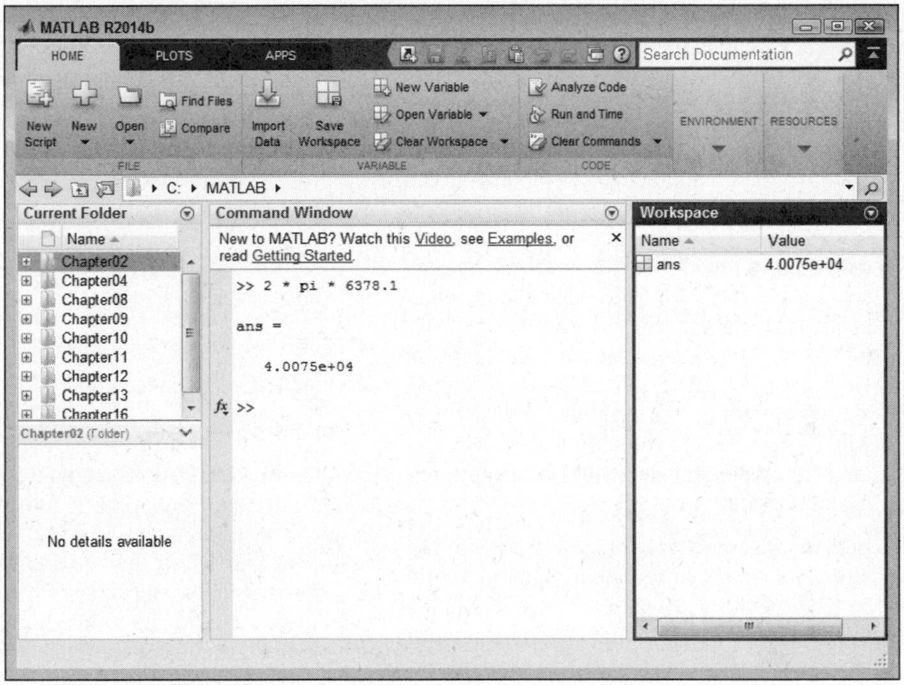

Figure 3-2: Any formula you enter changes the content of ans.

- ✔ Click Copy on the QAT.
- ✔ Right-click the highlighted text and choose Copy from the context menu.
- ✔ Rely on a platform-specific method of copying the text, such as pressing Ctrl+C on Windows.

When you have the text on the Clipboard, you can paste it wherever you want. If you want to paste it somewhere in MATLAB, click wherever you want to put the text, such as after the prompt. Use one of these methods to paste the text:

- ✔ Click Paste on the QAT.
- ✔ Right click the insertion point and choose Paste from the context menu.
- ✔ Rely on a platform-specific method of pasting text, such as pressing Ctrl+V on Windows.

Understanding integer and floating-point values

Throughout the book, you see the terms *integer* and *floating point*. These two terms describe kinds of numbers. When most people look at 3 and 3.0, they see the same number: the value three. The computer, however, sees two different numbers. The first is an integer — a number without a decimal portion. The second is a floating-point value — a number that has a decimal portion, even if it's a whole number.

You see these two terms often in this book because the computer works with and stores integer values differently from floating-point values. How the computer interacts differently with them is not important — you just need to know that it does. MATLAB does a great job of hiding the differences from view unless the difference becomes important for some reason, such as when you want to perform integer math — in which you want to work with only whole numbers. For example, 4 divided by 3 is equal to 1 with a remainder of 1 when performing integer math.

Humans also don't pay much attention to the size of a number. Again, the computer must do so because it has to allocate memory to hold the number — and larger numbers require more memory. So, not only do you need to consider the kind of number but also the size of the number when performing some tasks.

Finally, the computer must also consider whether a number has a sign associated with it. The sign takes up part of the memory used to store the number. If you don't need to store a sign, the computer can use that memory to store additional number information. With all these points in mind, here are the kinds of numbers that MATLAB understands:

- ✔ **double:** 64-bit floating-point double precision
- ✔ **single:** 32-bit floating-point double precision
- ✔ **int8:** 8-bit signed integer
- ✔ **int16:** 16-bit signed integer
- ✔ **int32:** 32-bit signed integer
- ✔ **int64:** 64-bit signed integer
- ✔ **uint8:** 8-bit signed integer
- ✔ **uint16:** 16-bit signed integer
- ✔ **uint32:** 32-bit signed integer
- ✔ **uint64:** 64-bit signed integer

Sometimes MATLAB won't know what you mean when you type 3. A value of 3 could be any kind of number. (MATLAB defaults to assuming that all values are doubles unless you specify otherwise.) To specify the kind of number you mean, you enter the type name and place the value in parentheses. For example, double(3) is a 64-bit floating-point number, but int32(3) is a 32-bit signed integer form of the same number.

Changing the Command window formatting

The Command window provides the means necessary to change the output formatting. For example, if you don't want the extra space between lines that MATLAB provides by default, you can type **format compact** and press Enter to get rid of it. In fact, try typing that command now. When you type **format**

compact and press Enter, you don't see any output. However, the next formula you type shows the difference. Type **2 + 2** and press Enter. You see that the extra spaces between lines are gone, as shown in Figure 3-3.

Figure 3-3:
Modify the appearance of the Command window using format commands.

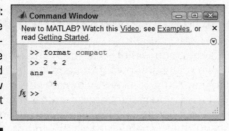

MATLAB provides a number of format commands. Each of them begins with the keyword format, followed by an additional instruction. Here is a list of the instructions you can type:

- ✔ **short:** All floating-point output has at least one whole number, a decimal point, and four decimal values, such as 4.2000.

- ✔ **long:** All floating-point output has at least one whole number, a decimal point, and 15 decimal values, such as 4.200000000000000.

- ✔ **shorte:** All floating-point output uses exponential format with four decimal places, such as 4.2000e+00.

- ✔ **longe:** All floating-point output uses exponential format with 15 decimal places, such as 4.200000000000000e+00.

- ✔ **shortg:** All output uses a short general format, such as 4.2, with five digits of space.

- ✔ **long:** All output uses a long general format, such as 4.2, with 15 digits of space.

- ✔ **shorteng:** All floating-point output uses exponential format with four decimal places and powers in groups of three, such as 4.2000e+000.

- ✔ **longeng:** All floating-point output uses exponential format with 14 decimal places and powers in groups of three, such as 4.20000000000000e+000.

- ✔ **hex:** All output is in hexadecimal format, such as 4010cccccccccccd.

- ✔ **+:** All output is evaluated for positive or negative values, so that the result contains just a + or - sign, such as + when using the formula 2 * 2.1.

- ✔ **bank:** All output provides two decimal places, even for integer calculations, such as 4.20.

✔ **rat:** All output is presented as a ratio of small integers, such as 21/5 for 4.2.

✔ **compact:** All output appears in single-spaced format.

✔ **loose:** All output appears in double-spaced format.

Suppressing Command window output

When performing most experiments, you want to see the result of your actions. However, sometimes you really don't want to keep seeing the results in the Command window when you can just as easily look in the Workspace window for the result. In these cases, you can follow a command with a semicolon (;) and the Command window output is suppressed. For example, try typing **2 + 2;** and pressing Enter (note the semicolon at the end of the command). You see output similar to that in Figure 3-4.

Figure 3-4:
Use a
semicolon
to hide the
results of an
action in the
Command
window.

Now look at the Workspace window. The results are shown there just as you would expect. This technique is often used when you have a complex set of formulas to type and you don't want to see the intermediate results or when working with large matrices. Of course, you also want to use this approach when you create scripts so that the script user isn't bombarded by the results that will appear as the script runs. Anytime you stop using the semicolon at the end of the command, you start seeing the results again.

Understanding the MATLAB Math Syntax

The MATLAB syntax is a set of rules that you use to tell MATLAB what to do. It's akin to learning another human language, except that the MATLAB syntax is significantly simpler than any human language. In order to communicate with MATLAB, you must understand its language, which is essentially a form of math. Because you already know math rules, you already know many

MATLAB rules as well. The following sections get you started with the basics that you use to build an understanding of the MATLAB language. You may be surprised to find that you already know some of these rules, and other rules are simply extensions of those rules.

Adding, subtracting, multiplying, and dividing

MATLAB is a math-based language, so it pays to review the basic rules for telling MATLAB how to perform basic math tasks. Of course, MATLAB performs the basic math functions:

- ✔ `+` or `plus()`: Adds two numbers. For example, you can use `3 + 4` or `plus(3, 4)` to obtain a result of `7`.

- ✔ `-` or `minus()`: Subtracts two numbers. For example, you can use `3 - 4` or `minus(3, 4)` to obtain a result of `-1`.

- ✔ `*` or `times()`: Multiplies two numbers. For example, you can use `3 * 4` or `times(3, 4)` to obtain a result of `12`.

- ✔ `/` or `rdivide()`: Performs right division, which is the form of division you likely learned in school. For example, you can use `3 / 4` or `rdivide(3, 4)` to obtain a result of `0.75`.

- ✔ `\` or `ldivide()`: Performs left division, which is also called "goes into" or, as you learned in third grade, "guzintas." You know (say this out loud), 5 "guzinta" 5 once, 5 "guzinta" 10 twice, 5 "guzinta" 15 three times, and so on. For example, you can use `3 \ 4` or `ldivide(3, 4)` to obtain a result of `1.3333`.

Most MATLAB operators are binary, which means that they work on two values. For example, `3 + 4` has two values: `3` and `4`. However, some operators are unary, which means that they work on just one value. Here are the basic unary operators:

- ✔ `+` or `uplus()`: Returns the unmodified content of a value or variable. For example, `+1` or `uplus(1)` is still equal to `1`.

- ✔ `-` or `uminus()`: Returns the negated content of a value or variable. For example, `-1` or `uminus(1)` returns `-1`. However, `--1` or `uminus(-1)` returns `1` (the negative of a negative is a positive).

In some cases, you don't want a floating-point result from division. To perform integer division, you have to use special functions — you can't just use operators for the simple reason that no operators are associated with these math tasks. Here are the functions associated with integer math:

✔ `idivide()`: Performs integer division. You supply two values or variables as input, along with an optional modifier that tells MATLAB how to perform rounding.

To use the `idivide()` function, you must specify that the input values are integers (see the "Understanding integer and floating-point values" sidebar in this chapter for details). For example, `idivide(int32(5), int32(3))` provides an output of 1. Here is a list of the modifiers you use to provide different rounding effects:

- `ceil`: Rounds toward positive infinity. For example, `idivide(int32(5), int32(3), 'ceil')` produces an output of 2 and `idivide(int32(5), int32(-3), 'ceil')` produces an output of –1.

- `fix`: Rounds toward zero. For example, `idivide(int32(5), int32(3), 'fix')` produces an output of 1 and `idivide(int32(5), int32(-3), 'fix')` produces an output of –1.

- `floor`: Rounds toward negative infinity. For example, `idivide(int32(5), int32(3), 'floor')` produces an output of 1 and `idivide(int32(5), int32(-3), 'floor')` produces a result of –2.

- `round`: Rounds to the nearest integer. For example, `idivide(int32(5), int32(3), 'round')` produces an output of 2 and `idivide(int32(5), int32(-3), 'round')` produces an output of –2.

✔ `mod()`: Obtains the modulus after division. For example, `mod(5, 3)` produces an output of 2 and `mod(5, -3)` produces an output of –1.

✔ `rem()`: Obtains the remainder after division. For example, `rem(5, 3)` produces an output of 2 and `rem(5, -3)` produces an output of –2.

Rounding can be an important feature of an application because it determines the approximate values the user sees. You can round any formula that you want to produce an integer output. Here are the rounding functions:

✔ `ceil()`: Rounds toward positive infinity. For example, `ceil(5 / 3)` produces an output of 2 and `ceil(5 / -3)` produces an output of –1.

✔ `fix()`: Rounds toward zero. For example, `fix(5 / 3)` produces an output of 1 and `fix(5 / -3)` produces an output of –1.

✔ `floor()`: Rounds toward negative infinity. For example, `floor(5 / 3)` produces an output of 1 and `floor(5 / -3)` produces an output of –2.

✔ `round()`: Rounds toward nearest integer. For example, `round(5 / 3)` produces an output of 2 and `round(5 / -3)` produces an output of –2.

Working with exponents

You use the caret (^) to raise a number to a particular power. MATLAB can handle negative, fractional, and complex number bases as exponents. Here are some examples of exponents:

- 10^3 = 1000
- 2^10 = 1024
- 2.5^2.5 = 9.8821
- 2^-4 = 0.0625
- 2^I = 0.7692 + 0.6390i
- i^I = 0.2079

Why we use the letter *E* (or *e*) for scientific notation

In the early days of computing, a display would use seven Light Emitting Diode (LED), or Liquid Crystal Display (LCD) segments to display numbers by turning particular segments on or off. Even today, many watches and clocks use this technique. The following figure shows how a seven-segment display works.

When designers made calculators that displayed scientific notation, they thought of the letter *E*, which reminds users that what follows is an exponent. They could also implement *E* using a seven- segment display, as shown here:

Then designers got lazy and instead of letting uppercase *E* mean scientific notation, they also let a lowercase *e* mean the same thing. In our modern age, designers can use all the pixels that various screens now employ to display the information without using the letter *E*. However, using *E* or *e* caught on, so now we have to use it. In addition, seven-segment displays are still commonly used in calculators, watches, clocks, and other devices.

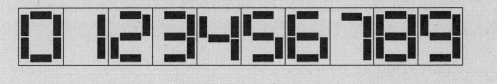

Organizing Your Storage Locker

Computers contain memory, much as your own brain contains memory. The computer's memory stores information that you create using MATLAB. Looking at memory as a kind of storage locker can be helpful. You open the door, put something inside and then close the door until you need the item again. When that happens, you simply open the door and take the item out. The idea of memory doesn't have to be complex or difficult to understand.

Whenever you tell MATLAB to store something in memory, you're using a *variable*. Developers use the term *variable* to indicate that the content of the memory isn't stable — it can change. The following sections tell you more about the MATLAB storage lockers called variables.

Using ans — the default storage locker

MATLAB always needs a place to store the output of any calculation you perform. For example, when you type 2 + 2 and press Enter, the output tells you that the value is 4. However, it more specifically tells you that ans = 4. MATLAB uses ans as a storage locker when you don't specify a specific storage locker to use.

MATLAB uses ans as a temporary storage locker. The content lasts only as long as you keep MATLAB open and you don't perform another calculation that requires ans to hold the output. If you need the result from a calculation for additional tasks, you must store the result in another variable.

Creating your own storage lockers

Whenever you need to use the result of a calculation in future calculations, you must create your own storage locker to hold the information; using the ans temporary variable just won't work. Fortunately, creating your own variables is straightforward. The following sections help you create your own variables that you can use for storing any MATLAB information you want.

Defining a valid variable name

A MATLAB variable name has certain requirements, just as naming other kinds of things must meet specific requirements. Here are the rules for creating a MATLAB variable:

✔ Start with a letter

✔ Add:

- Letters

- Digits

- Underscores

With this in mind, naming a variable 7Heaven doesn't work because this particular variable name begins with a number — and variables must begin with a letter. Likewise, Doug'sStuff doesn't work as a variable name because the apostrophe (') isn't allowed as part of a variable name. However, all the following variable names *do* work:

✔ MyVariable

✔ My_Variable

✔ My7Joys

In each case, the variable name begins with a letter and is followed by a letter, digit, or underscore. If you violate any of these rules, you see this error message:

```
Error: Unexpected MATLAB expression.
```

Always make variable names meaningful. Even though a variable named x is easy to type, remembering what x contains isn't so easy. A name such as CosOutput is much easier to remember because it has meaning. At least you know that it contains the output from a cosine calculation. The more meaningful you make the name, the easier it will be for you to later determine what a calculation does.

To create your own variable, type the variable name, an equal sign, and the value you want to assign to that variable. For example, to create a variable called MyName and assign it a value of Amy, you type **MyName = 'Amy'** and press Enter. (The single quotes show that Amy is a value [data], rather than another variable with the name of Amy.)

Understanding that variables are case sensitive

The "Entering information at the prompt" section, earlier in this chapter, discusses the need to type command and function names precisely as described in the MATLAB documentation because MATLAB is case sensitive. Variable names are also case sensitive, and this is one of the ways in which many users make mistakes when creating a script. The variable myVariable is different from MyVariable because the case is different.

Avoiding existing variable names

Avoiding the use of existing MATLAB names such as pi, i, j, sin, cos, log, and ans is essential. If you don't know whether a particular name is in use, you can type exist('*variable_name*') and press Enter. Try it now with pi. Type **exist('pi')** and press Enter. You see an output of 5, which means that the variable is in use. Now, type **exist('MyVariable')** and press Enter. The output is 0, which means that the variable doesn't exist.

MATLAB lets you create case-sensitive variations of existing variables. For example, type **Ans = 'Hello'** and press Enter. You see that the Workspace window now displays two variables, ans and Ans, as shown in Figure 3-5. Using a variable with the same name but different capitalization as an existing MATLAB variable will cause you problems. You're better off to simply avoid any existing term no matter how you capitalize it.

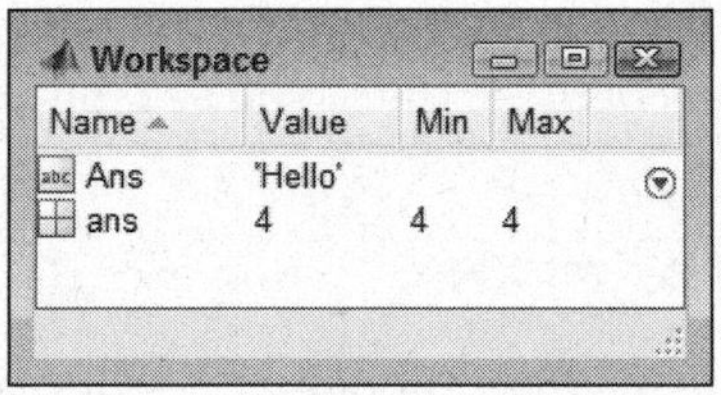

Figure 3-5:
Use unique names for your variables so that you can more easily avoid typing mistakes.

Workspace			
Name ▲	Value	Min	Max
Ans	'Hello'		
ans	4	4	4

Operating MATLAB as More Than a Calculator

It's time to take your first steps beyond using MATLAB as a simple calculator. The following sections help you get started using some of the MATLAB functions that you will eventually use to perform complex tasks.

Learning the truth

Determining whether something is true is an important part of performing most tasks. You determine the truth value of the information you receive almost automatically thousands of times a day. Computers can perform comparisons and report whether something is true (it does compare) or false (it doesn't compare). A man named George Boole (see `http://en.` `wikipedia.org/wiki/George_Boole`) created a method for quantifying the truth value of information using Boolean logic.

The basic idea is to ask the computer to perform a comparison of two variables. Depending on the values in those variables, the computer will say that it's either true that they compare or false that they compare. Table 3-1 spells out how Boolean logic works within MATLAB (where an output value of 1 means the statement is true and an output value of 0 means the statement is false).

Table 3-1	Relational Operators	
Meaning	*Operator*	*Example*
Less than	A < B	A=2; B=3; A==B ans = 1
Less than or equal to	A <= B	A=2; B=3; A==B ans = 1
Equal	A == B	A=2; B=3; A==B ans = 0
Greater than or equal to	A >= B	A=2; B=3; A==B ans = 0
Greater than	A > B	A=2; B=3; A==B ans = 0
Not equal	A ~= B	A=2; B=3; A==B ans = 1

It's essential to remember that one equal sign (=) is an assignment operator. It assigns the value you provide to the variable. Two equal signs (==) is an equality operator. This operator determines whether two variables contain the same value.

Using the built-in functions

Previous sections of this chapter introduce you to a number of MATLAB functions, but we have barely scratched the function surface. MATLAB has a lot of other functions, such as `sin()`, `cos()`, `tan()`, `asin()`, `acos()`, `atan()`, `log()`, and `exp()`. Many of these functions appear in other chapters of the book.

For an exhaustive list of functions, go to Appendix A. Yes, there really are that many. The appendix has brief descriptions of each function. Also, you can get additional information by typing `help('`*`function_name`*`')` and pressing Enter. Try it now. Type **help('sin')** and press Enter. You see output similar to that shown in Figure 3-6.

Command Window
New to MATLAB? Watch this Video, see Examples, or read Getting Started.
>> help('sin')
sin Sine of argument in radians.
sin(X) is the sine of the elements of X.
See also asin, sind.
Overloaded methods:
sym/sin
Reference page in Help browser
doc sin
fx >>

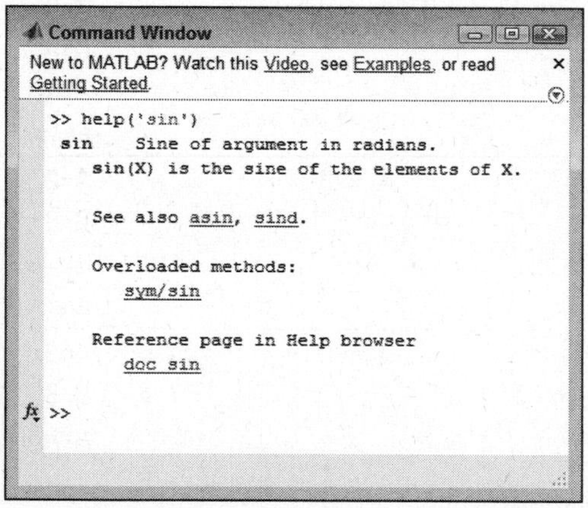

Figure 3-6: MATLAB makes it easy for you to learn more about functions you need.

Notice that the help screen contains links. Click any link to receive additional information about that topic.

Accessing the function browser

With all the functions that MATLAB provides, you might think it's impossible to discover what they are without a lot of memorization. Fortunately, help is closer than you might think. Look carefully at the Command window and you see an fx symbol in the border next to the prompt. Click the down arrow under the symbol and you see the dialog box shown in Figure 3-7. The official name of this dialog box is the Function Browser, and you use it to browse through categories of functions to track down the function you want.

You can also access the Function Browser using these techniques:

- Right-click the Command window and choose Function Browser from the context menu.
- Press Shift+F1.

Now that you have a better idea of what the Function Browser is, it's time to look at it in more detail. The following sections provide additional information on using the Function Browser.

Looking through the Function categories

The Function Browser is designed for you to easily drill down into a topic until you find precisely what you need. For example, when you click the Mathematics folder, you see a number of subcategories, such as Elementary Math, Linear Algebra, and Interpolation. When you click Elementary Math, you see yet more subcategories, such as Arithmetic, Trigonometry, and Polynomials. When you finally get to a list of functions, you see the fx symbol next to the entries, as shown in Figure 3-8.

Searching for a particular function

Sometimes you already have a good idea of what you want to find. In such a case, you can type all or part of a function name in the search bar at the top of the Function Browser window. For example, type **sin** to see all the functions that relate to working with sine, as shown in Figure 3-9.

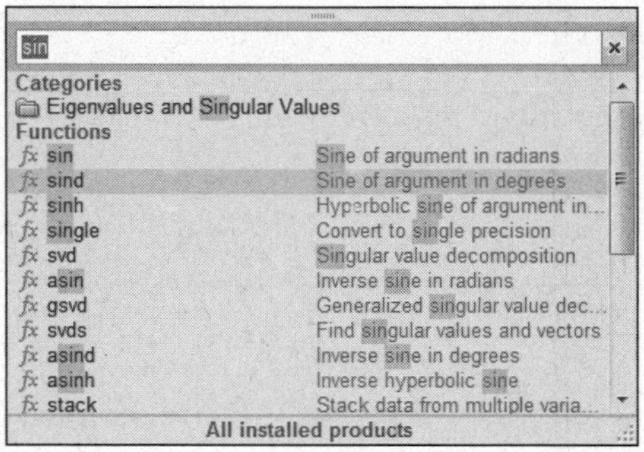

Figure 3-9:
Type search
terms to find
what you
need across
categories
quickly.

sin	×
Categories	
📁 Eigenvalues and Singular Values	
Functions	
fx sin	Sine of argument in radians
fx sind	Sine of argument in degrees
fx sinh	Hyperbolic sine of argument in...
fx single	Convert to single precision
fx svd	Singular value decomposition
fx asin	Inverse sine in radians
fx gsvd	Generalized singular value dec...
fx svds	Find singular values and vectors
fx asind	Inverse sine in degrees
fx asinh	Inverse hyperbolic sine
fx stack	Stack data from multiple varia...
All installed products	

Recovering from Mistakes

Everyone makes mistakes. You might think that experts don't make mistakes, but any expert who says so definitely isn't an expert. Making mistakes is part of the learning process. It's also part of the discovery process. If you want to do anything important with MATLAB, you're going to make mistakes. The following sections help you understand what to do when mistakes happen.

Understanding the MATLAB error messages

MATLAB tries to be helpful when you make mistakes. It doesn't always succeed, and you may not always understand the message, but it does try. In most cases, you see an error message that provides enough information for you to at least get started in finding the mistake. For example, if you try to use the `clc` command but type it in uppercase, you get

```
Undefined function or variable 'CLC'.
```

The error message is enough to get you looking for a solution to the problem, even when the problem isn't completely clear. In some cases, MATLAB even provides the correct command for you. All you have to do is press Enter and it executes.

Some errors are a little harder to figure out than others. For example, Figure 3-10 shows what happens when you try to use `idivide()` without specifying that the inputs are integers.

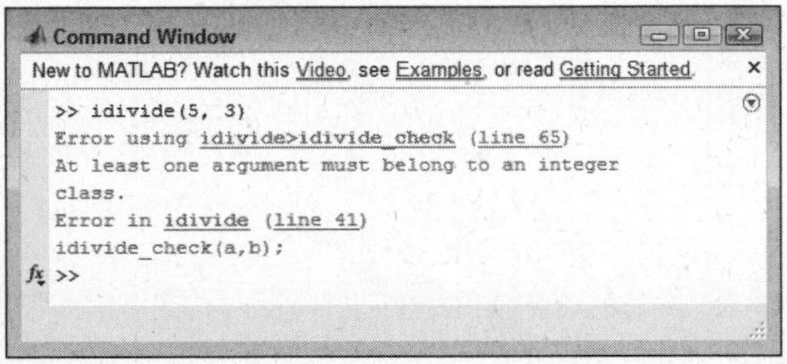

Figure 3-10:
Some error
messages
are a bit
complex.

```
Command Window
New to MATLAB? Watch this Video, see Examples, or read Getting Started.    ×
    >> idivide(5, 3)
    Error using idivide>idivide_check (line 65)
    At least one argument must belong to an integer
    class.
    Error in idivide (line 41)
    idivide_check(a,b);
fx >>
```

REMEMBER

In this case, you can ignore the links and what looks like gobbledygook. Focus on the second line. It tells you that one of the arguments must belong to the integer class. (Remember that the default is to assume that all numbers are doubles.) It's really saying that you need integer values as input to `idivide()`. When you get past the odd bits of information, you can more easily figure out how to fix the problem.

Stopping MATLAB when it hangs

Most of the time, MATLAB is extremely forgiving. You can make absolutely horrid mistakes, and MATLAB simply provides what it considers a helpful message without destroying anything. However, at times MATLAB has to chew on a bit of code for a while before it discovers the error, such as when you're working with a really large array. You can tell that MATLAB is working because the status bar shows Busy rather than Ready. In this case, you can talk to your buddy in the next cubicle, get a cup of coffee and read a good book, or press Ctrl+C to stop MATLAB from going any further.

REMEMBER

Pressing Ctrl+C always stops MATLAB from performing any additional processing. The status bar indicates Ready as soon as the processing is completely stopped. It's important that you do not use this option unless you really need to do so because MATLAB truly does stop right in the middle of what it's doing, which means that whatever you were doing is in an uncertain state. It's good to know that the option exists, though.

Getting Help

Just as everyone makes mistakes, so everyone needs help from time to time. Even the experts can't remember everything that MATLAB does, and notes go only so far in jogging the memory. So, when you need help, don't feel as though you're the only one seeking it. Most of the MATLAB-specific help information appears in the Resources group of the Toolstrip's Home tab, as shown in Figure 3-11. The following sections provide additional information on ways to obtain help.

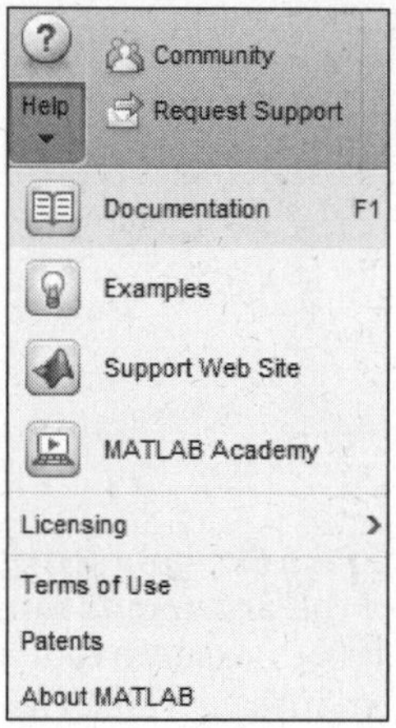

Figure 3-11:
The Resources group makes it easy to locate the help you need.

Exploring the documentation

The MATLAB documentation is complex and sometimes easy to get lost in when you look through it. Here are some ways to make the task a bit easier:

- ✔ Choose Help⇨Documentation in the Resources group of the Toolstrip's Home tab when you want to explore the documentation in general — simply as a means of learning more about MATLAB.

  If you want to find something a bit more specific, you can always type search terms in the search bar that appears at the top of the Help window. As you type, MATLAB displays corresponding topics in a manner that helps you narrow the focus of your search.

- ✔ Type **help('help_*topic*')** and press Enter in the Command window to obtain help about a specific help topic.

- ✔ Highlight a keyword or function name and press F1 to obtain help on that specific topic.

- ✔ Click links as provided in help messages, error messages, or other MATLAB output.

Working through the examples

You can access the examples that MATLAB provides by choosing Help⇨ Examples in the Resources group of the Toolstrip's Home tab. See Chapter 2 for information about how the examples work.

Relying on peer support

Peer support depends on other MATLAB users to help you. Because some other user has likely already encountered the problem you're having, peer support is a great option. To access peer support, click the Community icon in the Resources group of the Toolstrip's Home tab. You see your browser open to the MATLAB Central site, shown in Figure 3-12.

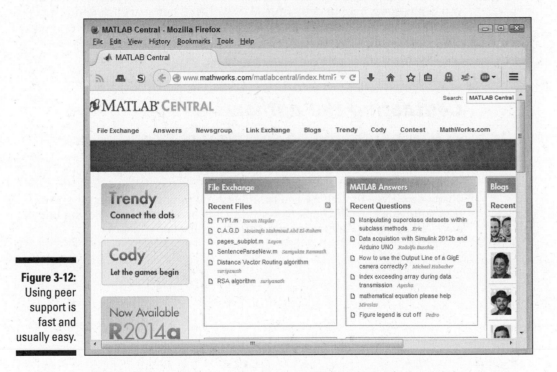

Figure 3-12: Using peer support is fast and usually easy.

The content on MATLAB Central changes regularly, but you normally see links for exchanging files, answers to common questions directly from MATLAB, blogs, and a number of other useful information areas. MATLAB Central is actually the best place to find what you need. Of course, you can always search the remainder of the Internet when MATLAB Central fails to provide precisely what you want.

Obtaining training

MATLAB offers courses in specific disciplines and in general usage. When you choose Help⇨Training in the Resources group of the Ribbon's Home tab, MATLAB takes you to a site where you can sign up for courses. The courses come in both traditional classroom form and in online format. The online format courses are further divided into those that are led by instructors and those that are self-paced.

Requesting support from MathWorks

When you have a really tough problem, one that defies any other solution, you can request help directly from MathWorks. When you click Request Support in the Resources group of the Toolstrip's Home tab, you see a login dialog box. Simply provide your email address and MathWorks password; then follow the prompts to obtain help from MathWorks.

Contacting the authors

When you have a question about the book, make sure that you contact either John (John@JohnMuellerBooks.com) or Jim (jsiz@tjc.edu). You can also find useful MATLAB information on this book's blog at http://blog.johnmuellerbooks.com/category/technical/matlab-for-dummies/. Even though we can't troubleshoot your application for you or take your exam, we're happy to help with any book-specific topic you might have. We're here to help you have a great reading experience!

Chapter 4

Starting, Storing, and Saving MATLAB Files

. . . .

In This Chapter
▶ Understanding the MATLAB file structure
▶ Working with MATLAB files
▶ Storing data on disk

. . . .

Computers have two kinds of storage bins: temporary memory in RAM and permanent memory on a storage device such as a hard drive. In order to make anything you create using MATLAB permanent, you must place it on the hard drive. Unfortunately, hard drives are huge, and if you want to find the data again later, you need to know where you placed the information. That's why knowing something about the MATLAB file structure is important — because you use it to find a place to store your data and to recover that data later.

Data is stored in files, while folders are used to organize the data. To load your data into MATLAB, you must first find the right folder, open it, and then open the file. It works much the same as a filing cabinet. As long as the drawer is closed and the file folder remains inside, the data is inaccessible. Note as well that some of your data may be in the wrong format. When data formatting is a problem, you need to import the data into MATLAB so that MATLAB can make use of it. The same holds true of other applications. When you want to use your MATLAB data with another application, you export it to that application.

The final section of this chapter discusses how to save your work for later use. The act of saving your work moves the data from temporary storage in RAM to permanent storage on the hard drive in a file. Later, when you need to access the data again, you open the file, which moves it from the hard drive into RAM. There is nothing mystical about this process. You perform the same sorts of tasks in the real world every day. Just think about how you use files in a filing cabinet the next time you open one.

Examining MATLAB's File Structure

To keep your data permanently, you must store it on disk. Of course, you could just store it anywhere, but then finding it later would be intensely difficult. In fact, given the size of today's hard drives, you might well retire before you find the data again. So, relying on some organized method of storing your information is important.

Applications also rely on specific file types when storing information. The main reason for using a specific file type is to allow the application to recognize its data among all the other data on your drive. Imagine the chaos if every application used the `.txt` file extension for every file. Not only would you become confused but the computer would become confused as well. In addition, using specific file types lets you know what sort of data the file contains.

MATLAB lets you identify the particular kind of information a file holds through the use of unique file extensions. For example, scripts and functions are stored in files with an `.m` extension, variables are stored in files with a `.mat` extension, and plots are stored in files with a `.fig` extension. In addition, you can organize your data using a file structure. You can perform all these management tasks from within MATLAB using either the application's GUI or commands. The following sections tell you how all these features work.

Understanding the MATLAB files and what they do

MATLAB provides specific file types for specific needs. The following list tells you about the MATLAB file types and describes their uses:

- ✔ `.fig`: Provides access to any plots or other graphics you create. Keep in mind that the file contains all the information required to reconstruct the graphic, but does not contain the graphic itself. This approach means that your image is accessible on any platform that MATLAB supports.

  A lot of people have asked whether they can access `.fig` files without necessarily having to display the graphic image itself. It turns out that `.fig` files are actually `.mat` files in disguise. The file format is the same (even though the content between the two file types differs). The article at `http://undocumentedmatlab.com/blog/fig-files-format` describes how you can access `.fig` files in text format so that you can see what they contain without seeing the associated graphic.

- ✔ `.m`: Holds a MATLAB script. This is a platform-independent file, so you can use the same scripts on whatever platform you're working on at the time. This file also allows you to create a script on one platform and share it with others, even when they use a different platform than you do. MATLAB script files are always written using the MATLAB language.

✔ `.mat`: Provides access to any data you saved on disk. Opening this file starts the Import Wizard to load the data into the MATLAB workspace.

✔ `.mdl`: Contains an older version of a Simulink model (see `.slx` below for details on the Simulink model). MATLAB recommends updating these files to the `.slx` format using the procedure at `http://www.mathworks.com/help/simulink/examples/converting-from-mdl-to-slx-model-file-format-in-a-simulink-project.html`.

✔ `.mex*`: Contains compiled executable code that extends MATLAB functionality in some manner. You execute these files just as you would a script program. The original code is written in either FORTRAN or C++ and then compiled for a specific platform. Each platform has a unique extension associated with it, as shown in the following list:

- `.mexa64`: Linux
- `.mexmaci64`: Mac OS X
- `.mexw32`: 32-bit Windows
- `.mexw64`: 64-bit Windows

✔ `.p`: Performs the same task as an `.m` file, except the content is protected from edits by anyone else. This feature lets you distribute your scripts to other people without fear of giving away programming techniques or trade secrets.

✔ `.slx`: Contains a Simulink model. Simulink is an add-on product for MATLAB that provides a block diagram environment for performing simulations. You can read more about this product at `http://www.mathworks.com/help/simulink/gs/product-description.html`. This book doesn't discuss the Simulink add-on because it's an advanced product used for higher-end needs.

Exploring folders with the GUI

The GUI method of working with folders in MATLAB requires the Current Folder window shown in Figure 4-1. (To display this window, choose Layout⇨Current Folder in the Environment group of the Toolstrip's Home tab.) In this case, the Current Folder toolbar appears at the top of the Current Folder window. You can also place it below the Toolstrip by choosing Layout⇨Current Folder Toolbar⇨Below Toolstrip in the Environment group of the Home tab. (The screenshots in the rest of the book assume you have selected the Inside Current Folder option.)

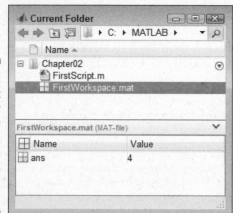

Figure 4-1: The Current Folder window provides GUI access to the MATLAB folders.

The Current Folder toolbar shows the current folder that the Current Folder window displays. To change locations, simply type the new location in the field provided. You can also select a location by clicking the right-pointing arrow next to each level, as shown in Figure 4-2. The arrow changes to a down-pointing arrow with a list of destinations below it. Clicking the magnifying glass icon in the field turns it into a Search field where you can choose the kind of file you want to find.

Figure 4-2: You can choose new locations by clicking the right-pointing arrow.

The Current Folder toolbar also includes four buttons. Each of these buttons helps you move to another location on the hard drive as follows:

- **Back:** Moves the location back one position in the history list. The history list is a list that is maintained by MATLAB that tracks the locations you've visited.

- **Forward:** Moves the location forward one position in the history list.

✔ **Up One Level:** Moves the location up to the parent folder.

✔ **Browse for Folder:** Displays a Select New Folder dialog box that you can then use to find another location on the hard drive. (See Figure 4-3.) After you find the folder, highlight its entry and click Select Folder to select it.

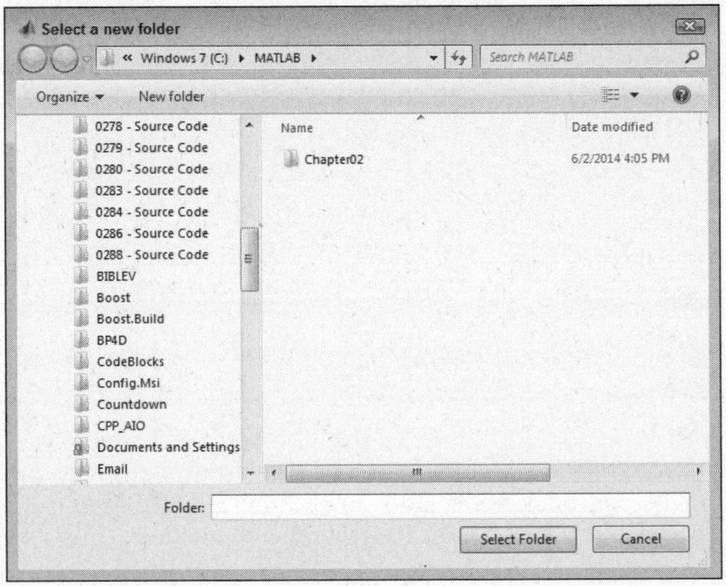

Figure 4-3: The Select New Folder dialog box helps you find other locations on the hard drive.

The Current Folder window provides access to all the folders that you've set up for organizational purposes. In this case, you see the Chapter02 subfolder (child folder) of the C:\MATLAB folder. The Chapter02 folder contains two files. When you right-click the Chapter02 folder entry, you see a number of commands on a context menu like the one shown in Figure 4-4.

Note that not all the entries on the context menu have to do with exploring folders or managing them from a file structure perspective. The following list focuses on those commands that *do* help you manage the file structure.

✔ **Open:** Opens the folder so that it becomes the current folder in the Current Folder toolbar.

✔ **Show in Explorer (Windows only):** Opens a copy of Windows Explorer so that you can interact with the folder using this Windows tool.

✔ **Create Zip File:** Creates a new .zip file that contains the compressed content of the folder. This feature makes sending the folder to someone else easier.

✔ **Rename:** Changes the name of the folder.

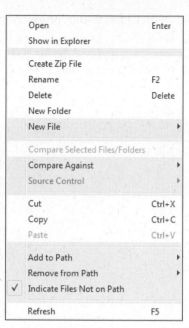

✔ **Delete:** Removes the folder and its content from the hard drive. Depending on how you have your system configured, this option could permanently destroy any data found in the folder, so use it with care.

✔ **New Folder:** Creates a new child folder within the selected folder.

✔ **New File:** Creates a new file within the folder. You can choose from these types of files:

- Script

- Function

- Example

- Class

- Zip File

✔ **Compare Against:** Matches the content of the selected folder against another folder and tells you about the differences.

✔ **Cut:** Marks the folder for removal from the hard drive and places a copy on the Clipboard. The folder is removed from its current location when you paste the copy in its new location.

✔ **Copy:** Places a copy of the folder and its content on the Clipboard so that you can paste copies of it in other locations.

✔ **Paste:** Places a copy of a folder and its content as found on the Clipboard in the location you indicate.

> ✔ **Refresh:** Verifies that the presentation of folders and files in the Current Folder window matches the actual content on the hard drive. Another application may have made modifications to the hard drive content, and with this command you can synchronize MATLAB with the physical device.

Exploring folders with commands

Many people prefer not to use the mouse. In this case, you can duplicate most of the GUI methods of interacting with the current folder using keyboard commands. The results of typing a command and pressing Enter appear in the Command window. To see how this feature works, try the following steps. (Your folder structure may not look precisely like the one in the book, but you should see appropriate changes as you type the commands.)

1. **Type** cd \MATLAB **and press Enter.**

    The Current Folder window changes to show the folder used for the book, as shown in Figure 4-5. (You may have to change the actual folder information to match your system if you chose not to create the directory structure described in earlier chapters.)

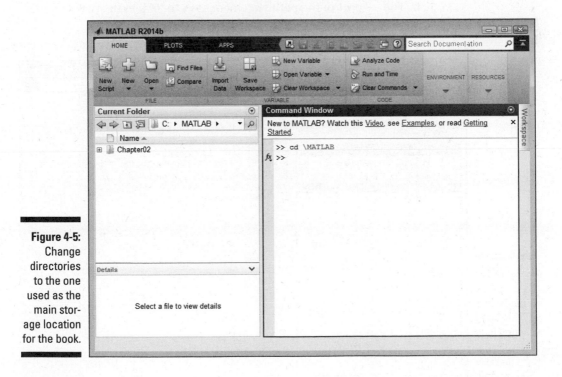

Figure 4-5:
Change directories to the one used as the main storage location for the book.

TIP

Even though you can't see it in this black-and-white book, MATLAB does provide color coding to make it easier for you to work with commands. Notice that the command portion of a command is in black lettering, while the argument part of the command is in purple lettering. The use of color coding helps you better see the commands and how they're structured.

2. **Type** mkdir Chapter04 **and press Enter.**

MATLAB creates a new folder to hold the materials for this chapter, as shown in Figure 4-6. Notice that you don't include a backslash (or slash) when creating a child directory for the current directory.

3. **Type** cd Chapter04 **and press Enter.**

The directory changes to the one used for this chapter. Notice (again) that you don't include a backslash (or slash) when moving to a subdirectory of the current directory.

4. **Type** copyfile ..\Chapter02\FirstScript.m **and press Enter.**

You see the copied file appear in the folder, as shown in Figure 4-7.

   a. The `copyfile` command provides the functionality needed to copy a file.

   b. The `..` part of the path statement says to look in the parent folder, which is \MATLAB.

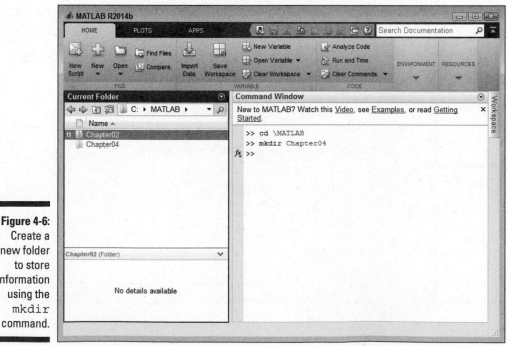

Figure 4-6: Create a new folder to store information using the `mkdir` command.

    c. The Chapter02 part of the path says to look in the Chapter02 subdirectory, which equates to \MATLAB\Chapter02.

    d. The FirstScript.m part of the path is the name of the file you want to copy to the current folder.

5. Type exist FirstScript.m **and press Enter.**

The command used in this case has a number of parts to it:

MATLAB provides an output value of 2, which means the file exists. This final step helps you validate that the previous steps all worked as intended. If one of the previous steps had gone wrong, you'd see a failure indicator, such as an error message or a different output value (as shown in the next step), with this step.

6. Type exist MyScript.m **and press Enter.**

In this case, the output value of 0 tells you that MyScript.m doesn't exist, as shown in Figure 4-8. The procedure didn't tell you to create MyScript.m, so this output is completely expected.

Now that you can see how the commands work, it's time to look at a command list. The following list contains an overview of the most commonly used file and folder management commands. (You can get detailed information at http://www.mathworks.com/help/matlab/file-operations.html.)

✔ cd: Changes directories to another location.

✔ copyfile: Copies the specified file or folder to another location.

✔ delete: Removes the specified file or object.

✔ dir: Outputs a list of the folder contents.

✔ exist: Determines whether a variable, function, folder, or class exists.

✔ fileattrib: Displays the file or directory attributes (such as whether the user can read or write to the file) when used without attribute arguments. Sets the file or directory attributes when used with arguments.

✔ isdir: Determines whether the input is a folder.

✔ ls: Outputs a list of the folder contents.

✔ mkdir: Creates a new directory.

✔ movefile: Moves the specified file or folder to another location.

✔ open: Opens the specified file using the default application. (Some files can be opened using multiple applications.)

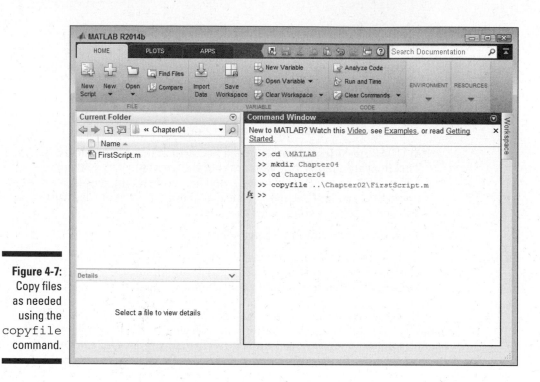

Figure 4-7:
Copy files
as needed
using the
`copyfile`
command.

✔ `pwd`: Displays the current path information, including the drive letter.

✔ `recycle`: Determines whether deleted files or folders are moved to the recycle bin.

✔ `rmdir`: Deletes the specified directory.

✔ `type`: Outputs the content of the specified file as text.

Some commands, such as `type`, can be combined with other commands, such as `disp`, to create well-formatted output. The `disp` command displays text, variables, or arrays. You discover how to use it later in the book (starting with Chapter 8). The point is that you sometimes combine commands to obtain a desired output.

✔ `visdiff`: Performs a comparison of two files of the following types:

- Text
- MAT-Files
- Binary
- Zip
- Folders

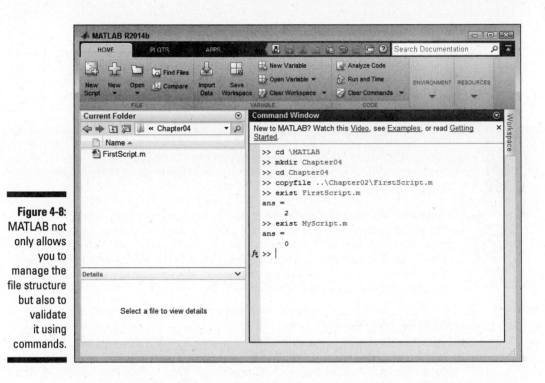

Figure 4-8: MATLAB not only allows you to manage the file structure but also to validate it using commands.

✔ what: Provides a categorized listing of MATLAB-specific files in the current directory. For example, if the current directory contains any files with an .m extension, you see them listed in the MATLAB code files category.

✔ which: Helps locate files and functions based on filename, function name path, or other criteria.

✔ winopen: Used only with Windows; opens the specified file using the default application. (Some files can be opened using multiple applications.)

Working with files in MATLAB

Folders provide organization, but files are what hold your data. Working with files is an essential part of learning to work with MATLAB. After all, if you can't find your data, you can't do anything with it. Managing the data to ensure that it remains safe, secure, and reliably accessible is important. The following sections describe how to perform common tasks with files.

Using the right-click to your advantage

Every file and folder shown in the Current Folder window has a context menu associated with it. A context menu always displays just the actions that you can perform with that file or folder. By right-clicking various files and folders, you see the context menu and might discover new tasks that you can perform with the file or folder you highlighted.

Right-clicking a file or folder can never damage it. The only time you might damage the file or folder is if you select an item from the context menu. To close the context menu after you view it, click in an empty area outside the context menu.

Depending on your platform, you may also see shortcut keys when viewing the context menu. For example, when working with Windows, you can highlight a file and press Ctrl+C to copy it to the Clipboard — all without using the context menu at all. Pasting is just as easy: Select the folder you want to use to store the file and press Ctrl+V. As mentioned, these shortcut keys are platform specific, which is why they aren't used in the book.

Copying and pasting

Copying and pasting creates a copy of an existing data file and places that copy in another location. You use this process in a number of ways. For example, you might want to make a backup of your data before you modify it, share the data with a friend, or place the data on removable media so that you can take it home and work on it. Even though the following steps use a specific file and locations, you can easily use them for any file you want to copy and paste and with any location. In this case, you copy `FirstWorkspace.mat` found in the `Chapter02` folder to the `Chapter04` folder.

1. **Open the `\MATLAB\Chapter02` folder in the Current Folder window.**

   You see two files: `FirstScript.m` and `FirstWorkspace.mat`, as shown in Figure 4-9. Note that your Current Folder window might not be arranged precisely the same as the one shown in Figure 4-9.

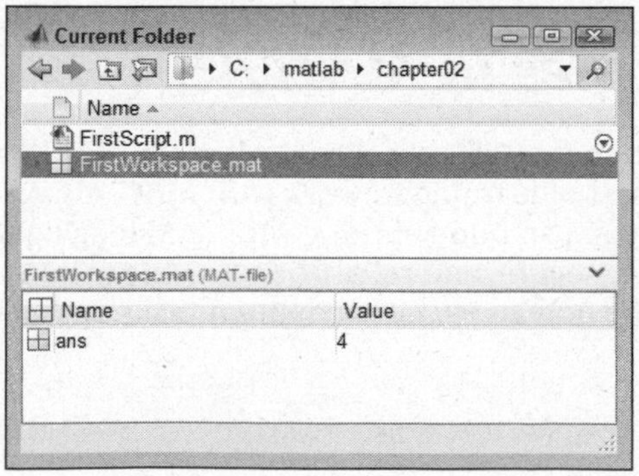

Figure 4-9: The Chapter02 folder contains two files.

2. **Right-click** FirstWorkspace.mat **and choose Copy from the context menu.**

   This action copies the file onto the Clipboard. You won't actually see anything happen in the window.

3. **Click the Up One Level button in the Current Folder toolbar.**

   You return to the \MATLAB folder. This folder should show two subdirectories: Chapter02 and Chapter04, as shown in Figure 4-10. If you don't see both subdirectories, make sure to create the Chapter04 subdirectory using the steps found in the "Exploring folders with commands" section, earlier in this chapter.

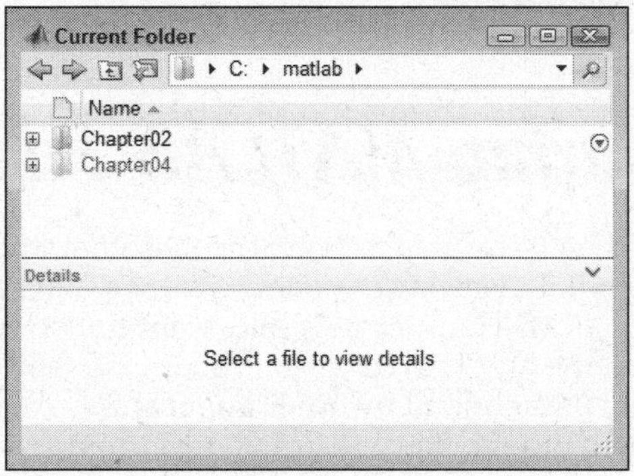

Figure 4-10: The MATLAB folder should contain two sub-directories.

> **Current Folder**
>
> ◁ ▷ ▣ ▨ ▮ ▸ C: ▸ matlab ▸
>
> 🗋 Name ▲
> ⊞ 📁 Chapter02
> ⊞ 📁 Chapter04
>
> Details
>
> Select a file to view details

Even though the screenshot in the book doesn't show it, the Chapter02 subdirectory is darker than the Chapter04 subdirectory. The reason for this difference is that the Chapter02 subdirectory is on your MATLAB path, while the Chapter04 subdirectory isn't. To add Chapter04 to the path, right-click its entry and choose Add To Path⇨Selected Folders or Add To Path⇨Selected Folders and Subfolders from the context menu.

4. **Double-click the** Chapter04 **folder to open it.**

   This folder should contain a single existing file, FirstScript.m.

5. **Right-click anywhere within the folder area and choose Paste.**

   MATLAB copies the file to the new location for you. At this point, the Chapter04 folder should look precisely like the Chapter02 folder in Figure 4-9.

Cutting and pasting

The process for cutting and pasting a file is almost the same as copying and pasting it. (See the previous section for details.) The only difference is that you select Cut rather than Copy from the context menu. However, the results are slightly different. When you cut and paste a file, the file is actually moved from one location to another. You use Cut and Paste when you don't want to create multiple copies of a file and simply want to place the file in another location.

Dragging

Dragging a file or folder moves it from one location to another. All you need to do is click the file. While you hold the mouse button down, you drag the file to a new location. MATLAB moves the file to the location you specify.

If the location already has a file with that name, MATLAB displays a message asking whether you're sure you want to move the file. You must confirm the move before MATLAB performs the task. The new file replaces the existing file, so you could experience data loss.

Accessing and Sharing MATLAB Files

To make data useful, you need to be able to open the files containing it. Otherwise, there isn't any point in saving the data. Likewise, not all your colleagues will have a copy of MATLAB, or they may want to use a different application to interact with the MATLAB data. For you to use their data, you must be able to *import* data files created by other applications. When you want to share your data with others, you must *export* your data to files that are understood by other applications. MATLAB provides great support for both imported and exported data.

Opening

The fastest way to open any MATLAB file is to double-click its entry in the folder found in the Current Folder window. You can also right-click the entry and choose Open from the context menu. MATLAB automatically opens the file using a default application or method.

It's important to realize that MATLAB always uses a default application or method. Data files are sometimes associated with other applications. In addition, some data files can be opened in more than one way.

When you want to use an alternative method of opening a file, you must rely on the underlying platform. For example, when working with Windows, right-click the file and choose Show in Explorer from the context menu. A copy of Windows Explorer opens, and you can work with alternative applications in that copy. Right-click the file in Windows Explorer and choose one of the alternative applications shown in the Open With menu of the context menu. Figure 4-11 shows an example for `FirstScript.m` where you can open the file using multiple applications (with the default shown at the top of the list).

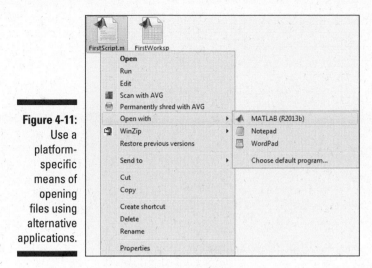

Figure 4-11:
Use a
platform-
specific
means of
opening
files using
alternative
applications.

MATLAB also uses different techniques for interacting with files when you work with commands. The default action for a .mat file is to load it into MATLAB, not open it. However, you can either load it or open it as needed. Here are the two commands you use (assuming that you want to work with FirstWorkspace.mat):

✔ open('FirstWorkspace.mat')

✔ load('FirstWorkspace.mat')

The first command actually opens the workspace so that you can see a result in the Command window. However, the results aren't loaded into the Workspace window as they normally would be if you double-clicked the file. To achieve this same effect, you must use the second command, which loads the workspace into MATLAB.

Importing

MATLAB makes importing whatever data you need from an external source easy. The following steps show you how:

1. **Click Import Data in the Variable group of the Home tab.**

   You see the Import Data dialog box, as shown in Figure 4-12. Notice that MATLAB defaults to showing every file it can import.

   If you find that the list of files is too long, you can click the Recognized Data Files drop-down list and choose just one of the common file types. The list displays just those files, making a selection easier.

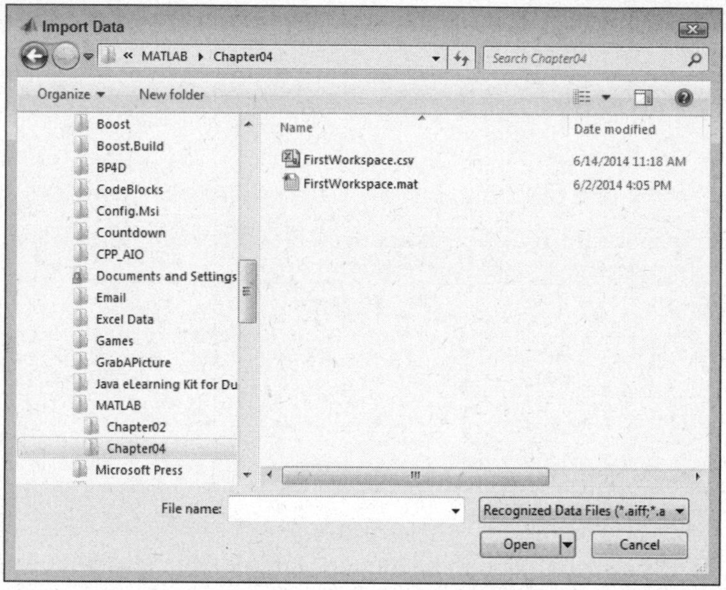

Figure 4-12:
The Import Data dialog box lets you choose which file to import.

2. Highlight the file you want to import and click Open.

MATLAB displays an Import dialog box that contains import information about the file, as shown in Figure 4-13. This dialog box contains settings that you use to import the data and ensure that it's useful in MATLAB. Figure 4-13 shows the settings for a comma-separated value (CSV) file, and the rest of the procedure assumes that you're working with such a file. However, the process is similar for other file types.

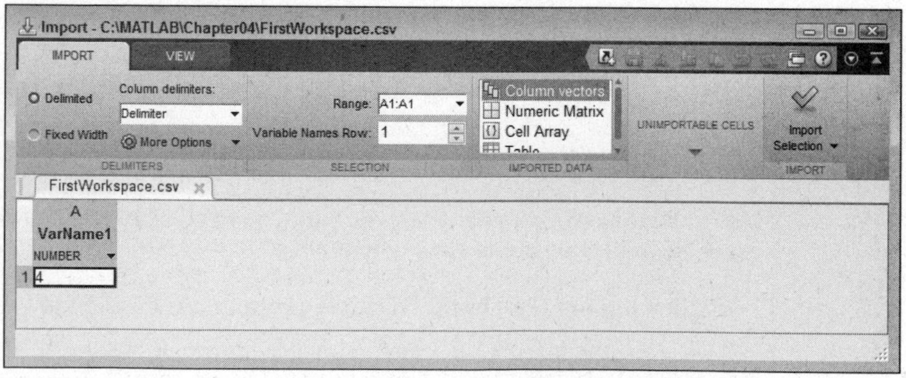

Figure 4-13:
The Import dialog box lets you tweak the import settings.

3. (Optional) Modify the settings as needed so that the data appears as it should appear in the Workspace window.

You can choose to limit the amount of data imported by changing the range. It's also possible to select a different delimiter (which changes how the data appears onscreen).

4. Verify that the Unimportable Cells group has no entries.

Cells that MATLAB can't import might reflect an error or simply mean that you have some settings wrong.

5. Click Import Selection.

MATLAB imports the data. As alternatives, you can also choose to generate a script or function based on the data, rather than actually import the data into the workspace.

6. Close the Import window.

You can read about the data formats that MATLAB can import at `http://www.mathworks.com/help/matlab/import_export/supported-file-formats.html`. This site also contains commands that you can use to import the files rather than relying on the GUI to do the work. However, the GUI is always faster and easier to use, so it's the recommended course.

Exporting

You rely on commands in order to export data from MATLAB. The list of data formats at `http://www.mathworks.com/help/matlab/import_export/supported-file-formats.html` includes commands in the Export column for each format that MATLAB supports.

Most of the commands work with a single variable. For example, if you want to export the information found in the `ans` variable to a CSV file, you type something like `csvwrite('FirstWorkspace.csv',ans)`, where `csvwrite()` is the function, `FirstWorkspace.csv` is the name of the file, and `ans` is the name of the variable you want to export.

Along with `csvwrite()`, the most commonly used export commands are `xlswrite()`, which creates an Excel file, and `dlmwrite()`, which creates a delimited file. Both of these commands work much the same as `csvwrite()`.

Some file formats require quite a bit of extra work. For example, to create an eXtensible Markup Language (XML) file, you must first build a document model for MATLAB to use. You can see the procedure for performing this task at `http://www.mathworks.com/help/matlab/ref/xmlwrite.html`.

Saving Your Work

An essential part of ending any session with MATLAB is saving your work. Otherwise, you could lose everything you've worked so hard to achieve. In fact, smart users save relatively often to avoid the power-failure penalty. How often you save depends on your personal work habits, the value of the work, and the potential need to use time and system resources efficiently. No matter how you save or when, the following sections help you get the job done.

Saving variables with the GUI

Although Chapter 2 does show you how to save the entire workspace, sometimes you need to save just one variable. You can perform this task using the GUI and the following steps:

1. **Right-click the variable that you want to save in the Workspace window and choose Save As from the context menu.**

   You see the Save to MAT-File dialog box, shown in Figure 4-14.

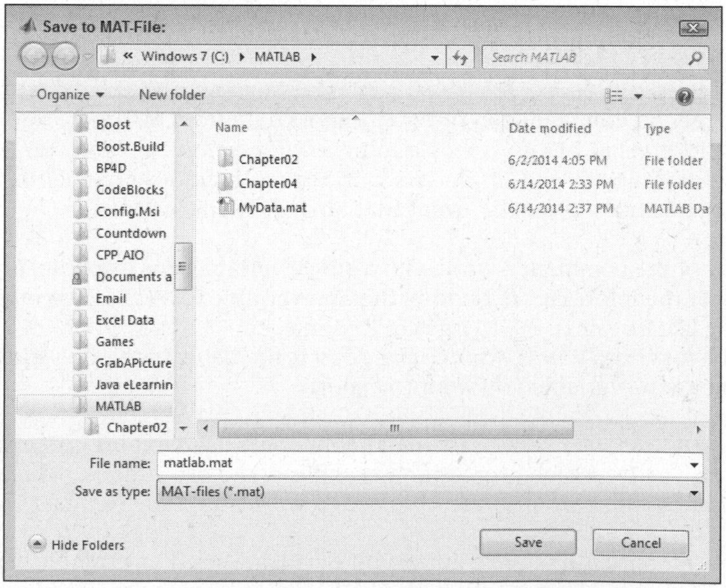

Figure 4-14:
Use the Save to MAT-File dialog box to save individual variables.

2. **Type a name for the file in the File Name field.**

   Choose something that will help you remember the purpose of the variable.

You can use the tree structure in the left pane to choose a different folder if you don't want to use the current folder to store the file containing the variable information.

3. Click Save.

MATLAB saves the variable to the file you choose.

Saving variables using commands

You can use commands to save your variables to disk. In fact, the command form is a little more flexible than the GUI. The basic command you use is `save('filename')`, where `filename` is the name of the file you want to use.

When you want to save specific variables, you must add a list of them after the filename. For example, `save('MyData.mat', 'ans')` would save a variable named `ans` to a file named `MyData.mat` in the current folder. You can include path information as part of the filename if you want to save the data in a different folder. For example, `save('C:\Temp\MyData.mat', 'ans')` would save the data in the `C:\Temp` folder. If you want to save multiple variables, simply create a comma-delimited list of them. To save `Var1` and `Var2` to `MyData.mat`, you type `save('MyData.mat', 'Var1', 'Var2')`.

These initial commands save the output in MATLAB format. However, you can also specify a format. The formats are listed at `http://www.mathworks.com/help/matlab/ref/save.html#inputarg_fmt`. For example, to save the previous variables in ASCII format, you type `save('MyData.txt', 'Var1', 'Var2', '-ASCII')`.

Saving commands with the GUI

You can't save commands that you type directly into the Command window using the GUI. What you do instead is save them using the Command History window. The "Saving a formula or command as a script" section of Chapter 2 describes how to save both formulas and commands.

Saving commands using commands

MATLAB does let you save commands to disk using a command: `diary`. A *diary* is simply an on-disk record of the commands that you type in the Command window. Later, you can review the file and edit it just as you would a script. The `diary` command actually has a number of forms, as follows:

- ✔ diary: Creates a diary file with the filename diary. Because this file has no extension, it isn't associated with anything. The output is ASCII, and you can open it with any text editor.

- ✔ diary('*filename*'): Creates a diary file that has the name *filename*. You can give the output file an .m extension, which means that you can open it as a script using the MATLAB editor. This approach is actually better than using diary by itself because the resulting file is easier to work with.

- ✔ diary off: Turns off recording of your commands so that they aren't recorded to the file. Setting the diary to off lets you experiment before committing commands that you don't want to the file on disk.

- ✔ diary on: Resumes recording of your commands.

Part II

Manipulating and Plotting Data in MATLAB

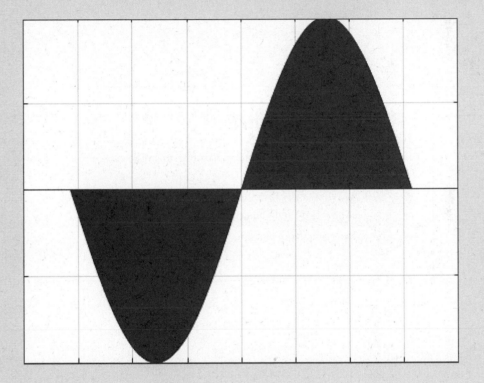

See an example of how you can plot formulas the easy way at `http://www.dummies.com/extras/matlab`.

In this part . . .

- ✔ See how to interact with vectors, matrices, and higher dimensions.

- ✔ Perform specific math tasks with vectors and matrices.

- ✔ Discover how to perform basic plotting tasks.

- ✔ Create more advanced plots that help better document your work.

Chapter 5

Embracing Vectors, Matrices, and Higher Dimensions

The previous chapters of this book introduce you to MATLAB and its interface. Starting in this chapter, you become immersed in math a little more serious than 2 + 2. Of course, in this "more serious" math, many problems revolve around vectors and matrices, so these are good topics to start with. This chapter helps you understand how MATLAB views both vectors and matrices and how to perform basic tasks with these structures. The chapter then takes you from two-dimensional matrices to matrices with three or more dimensions. All this material gives you a good idea of just how MATLAB can help you solve your vector and matrix problems.

Of course, you might still have questions. In fact, a single chapter of a book can't answer every question on this topic. That's why you also need to know how to obtain additional help. The last section of the chapter provides insights into how you can get additional help from MATLAB and force it to do more of your matrix work for you. (After all, MATLAB is there to serve your needs, not the other way around.)

Working with Vectors and Matrices

Vectors are simply a row of numbers. The length of this row has no limit and the numbers have no specific interval. *Matrices* are a two-dimensional table of numbers. Again, the size of this table has no limit (either in rows or columns) and the numbers have no specific interval. Both structures are well understood by mathematicians and engineers. They are used extensively by MATLAB to perform tasks that might otherwise require the use of complex structures not understood by these groups, which would unnecessarily complicate MATLAB usage.

The following sections describe how MATLAB uses vectors and matrices to make creating programs easier and demonstrates some of the ways in which MATLAB uses them. (Note that this chapter's discussion assumes that you're coming to the table with a basic understanding of linear algebra. If you find that you need to brush up on this particular area, check out the "Locating linear algebra resources online" sidebar.)

Understanding MATLAB's perspective of linear algebra

Linear algebra deals with vector spaces and linear mappings between those spaces. You use linear algebra when working with lines, planes, and subspaces and their intersections. When working with linear algebra, vectors are viewed as coordinates of points in space, and the algebra defines operations to perform on those points.

MATLAB divides linear algebra into these major areas:

- ✔ Matrix analysis
    - Matrix operations
    - Matrix decomposition
- ✔ Linear equations
- ✔ Eigenvalues
- ✔ Singular values
- ✔ Matrix functions
    - Logarithms
    - Exponentials
    - Factorization

This chapter doesn't cover absolutely every area, but you are exposed to enough linear algebra to perform most tasks effectively using MATLAB. As the book progresses, you see additional examples of how to make MATLAB perform tasks using linear algebra. Think of this chapter as a really good start toward the goal of making MATLAB perform linear algebra tasks for you at a level of speed and accuracy you couldn't achieve otherwise.

Locating linear algebra resources online

This chapter doesn't provide a tutorial on linear algebra. (We're assuming most of you would be bored by it anyhow because you're already math geniuses.) Of course, not everyone remembers that college course in linear algebra, and some things that you don't use every day are likely to be a little hard to remember. With this in mind, you might want to locate a linear algebra tutorial to jog your memory. Many good sources of information about linear algebra are available online.

One of the more interesting places to get some information about linear algebra is the Khan Academy at `https://www.khanacademy.org/math/linear-algebra`. Most of the information is relayed through videos, so you get the benefit of a classroom-like presentation. The presentations are short, for the most part — usually less than ten minutes — so you can watch segments as time presents. In addition, you can pick and choose among the videos to watch.

If all you really want is a quick brush up on linear algebra, you might not need something as time-consuming as what the Khan Academy provides. In that case, you might want to check out the linear algebra tutorial in four pages at `http://minireference.com/blog/linear-algebra-tutorial/`. A number of people using this resource complained that it went really fast. After reviewing it, we can report that the four pages are well done, but they really do assume that you need a light refresher and already know how to use linear algebra quite well.

A middle ground tutorial is found on Kardi Teknomo's Page at `http://people.revoledu.com/kardi/tutorial/LinearAlgebra/`. The interesting thing about this tutorial is that it's interactive. You get somewhat detailed text instruction and then get to try your new skills right there on the site. The act of reading the information and then practicing what you learn makes the information stick better.

The point is that you're likely to find a tutorial that meets your specific needs. You just need to invest a few minutes in trying out the various tutorials until you find one that fits your particular learning style. It simply isn't possible to provide such diversity in a single chapter of a book, so that's why the online resources are so important.

Entering data

Chapter 3 shows you how to import data from a spreadsheet or another data source. Of course, that's fine if you have a predefined data source. However, you'll often need to create your own data, so knowing how to type it yourself is important.

Think about how you use data when working with math: The data appears as a list of numbers or text. MATLAB uses a similar viewpoint. It also works with lists of numbers and text that you create through various methods. The following sections describe how to enter data as lists by using assorted techniques.

Entering values inside square brackets

The left square bracket, [, starts a list of numbers or text. The right square bracket,], ends a list. Each entry in a list is separated by a comma (,). To try this technique yourself, open MATLAB, type **b=[5, 6]** in the Command window, and press Enter. You see

```
b =
     5        6
```

The information is stored as a list of two numbers. Each number is treated as a separate value. Double-click b in the Workspace window and you see two separate entries, as shown in Figure 5-1. Notice that the Workspace window shows b as a 1 x 2 list in which the entries flow horizontally.

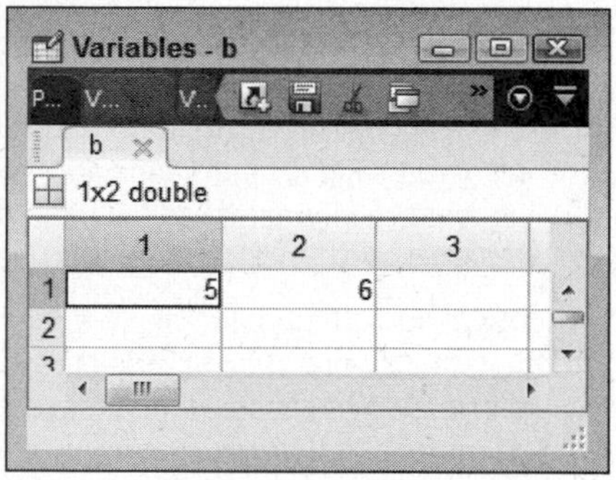

Figure 5-1:
Typing comma-separated numbers in square brackets produces a list of numbers.

You can type **format compact** and press Enter to save display space. If you want to clear space in the Command window for typing additional commands, type **clc** and press Enter. Chapter 3 provides additional details on configuring MATLAB output.

Starting a new line or row with the semicolon

The comma creates separate entries in the same row. You use the semicolon (;) to produce new rows. To try this technique yourself, type **e=[5; 6]** in the Command window and press Enter. You see

```
e =
     5
     6
```

As in the previous section, the information is stored as a list of two numbers. However, the arrangement of the numbers differs. Double-click e in the Workspace window and you see two separate entries, as shown in Figure 5-2.

Notice that the Workspace window shows e as a 2 x 1 list in which the entries flow vertically.

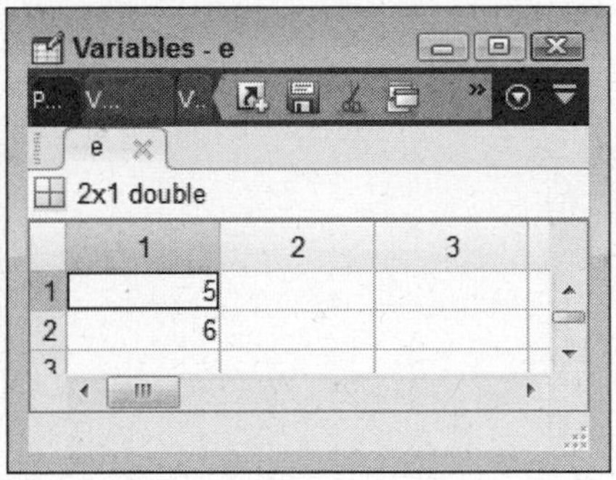

Figure 5-2:
Typing semicolon-separated numbers produces rows of values.

Figure 5-2: Typing semicolon-separated numbers produces rows of values.

Separating values with a comma or a semicolon

It's possible to create a matrix by combining commas and semicolons. The commas separate entries in the same row and the semicolons create new rows. To see this for yourself, type **a=[1, 2; 3, 4]** in the Command window and press Enter. You see

```
a =

     1     2
     3     4
```

Notice how the output looks like the linear algebra you're used to. MATLAB makes every effort to use a familiar interface when presenting information so that you don't have to think about how to interpret the data. If the output doesn't appear as you expect, it could be a sign that you didn't create the information you expected, either.

Finding dimensions of matrices with the Size column

Figures 5-1 and 5-2 show one way to obtain the size of a numeric list (it appears in the upper-left corner of the window). However, you can use an easier method. Right-click (Control+click on a Mac) the Workspace window column list and select Size from the context menu. (You can also right-click the title bar and select Choose Columns➪Size from the context menu.)

You may also find it helpful to display the minimum and maximum values for each entry. This information comes in handy when working with large vectors or matrices where the minimum and maximum values aren't obvious. To obtain this information, choose Columns➪Min, and then choose Columns➪max.

Depending on your computer screen, you may need to click and drag the Size, Min, and Max columns more to the left so that you can see them. You can also resize the window. Figure 5-3 shows the results of the entries you created in the previous sections.

Creating a range of values using a colon

Typing each value in a list manually would be time-consuming and error-prone because you'd eventually get bored doing it. Fortunately, you can use the colon (:) to enter ranges of numbers in MATLAB. The number on the left side of the colon specifies the start of the range, and the number on the right side of the colon specifies the end of the range. To see this for yourself, type **g=[5:10]** and press Enter. You see

```
g =
     5     6     7     8     9    10
```

Creating a range of values using linspace ()

Using the colon to create ranges has a problem. MATLAB assumes that the *step* (the interval between numbers) is 1. However, you may want the numbers separated by some other value. For example, you might want to see 11 values between the range of 5 and 10, instead of just 6.

The linspace() function solves this problem. You supply the starting value, the ending value, and the number of values you want to see between the starting and ending value. To see how linspace() works, type **g=linspace(5,10,11)** and press Enter. You see

```
g =
   Columns 1 through 5
      5.0000     5.5000     6.0000     6.5000     7.0000
   Columns 6 through 10
      7.5000     8.0000     8.5000     9.0000     9.5000
   Column 11
     10.0000
```

In this case, the step value is 0.5. Each number is 0.5 higher than the last, and there are 11 values in the output. The range is from 5 to 10, just as in the colon example in the previous section. In short, using `linspace()` is a little more flexible than using the colon, but using the colon requires less typing and is easier to remember.

Adding a step to the colon method

It turns out that you can also specify the step when using the colon method. However, in this case, you add the step between the beginning and ending of the range when defining the range. So, you type the beginning number, the step, and the ending number, all separated by colons. To try this method for yourself, type **g=[5:0.5:10]** and press Enter. You see

```
g =
  Columns 1 through 5
     5.0000     5.5000     6.0000     6.5000     7.0000
  Columns 6 through 10
     7.5000     8.0000     8.5000     9.0000     9.5000
  Column 11
    10.0000
```

This is precisely the same output as that of the `linspace()` example. However, when using this method, you specify the step directly, so you don't control the number of values you receive as output. When using the `linspace()` approach, you specify the number of values you receive as output, but MATLAB computes the step value for you. Each technique has advantages, so you need to use the one that makes sense for your particular need.

Transposing matrices with an apostrophe

Using the colon creates row vectors. However, sometimes you need a column vector instead. To create a column vector, you end the input with an apostrophe. To see how this works for yourself, type **h=[5:0.5:10]'** and press Enter. You see

```
h =
     5.0000
     5.5000
     6.0000
     6.5000
     7.0000
     7.5000
     8.0000
     8.5000
     9.0000
     9.5000
    10.0000
```

When you look at the Workspace window, you see that g is a 1 x 11 vector, while h is an 11 x 1 vector. The first entry is a row vector and the second is a column vector.

You can transpose matrices as well. The rows and columns change position. For example, earlier you typed a=[1,2;3,4], which produced

```
a =
     1     2
     3     4
```

To see how this matrix looks transposed, type **i=[1,2;3,4]'** and press Enter. You see

```
i =
     1     3
     2     4
```

Adding and Subtracting

Now that you know how to enter vectors and matrices in MATLAB, it's time to see how to perform math using them. Adding and subtracting is a good place to start.

The essential rule when adding and subtracting vectors and matrices is that they must be the same size. You can't add or subtract vectors or matrices of different sizes because MATLAB will display an error message. Use the following steps to see how to perform this task:

1. **Type** a=[1,2;3,4] **and press Enter.**

    You see

    ```
    a =
         1     2
         3     4
    ```

2. **Type** b=[5,6;7,8] **and press Enter.**

    You see

    ```
    b =
         5     6
         7     8
    ```

3. **Type** c = a + b **and press Enter.**

   This step adds matrix a to matrix b. You see

   ```
   c =
         6       8
        10      12
   ```

4. **Type** d = b - a **and press Enter.**

   This step subtracts matrix b from matrix a. You see

   ```
   d =
         4       4
         4       4
   ```

5. **Type** e=[1,2,3;4,5,6] **and press Enter.**

   You see

   ```
   e =
         1       2       3
         4       5       6
   ```

   If you attempt to add or subtract matrix e from either matrix a or matrix b, you see an error message. However, the following step tries to perform the task anyway.

6. **Type** f = e + a **and press Enter.**

   As expected, you see the following error message:

   ```
   Error using +
   Matrix dimensions must agree.
   ```

   The error messages differ a little between addition and subtraction, but the idea is the same. The matrices must be the same size in order to add or subtract them.

Understanding the Many Ways to Multiply and Divide

After adding and subtracting comes multiplication and division. MATLAB is just as adept in meeting this need as it is in every other area. The following sections describe the many ways in which you can use multiplication and division in MATLAB.

Performing scalar multiplication and division

A *scalar* is just technobabble for ordinary numbers. When you multiply ordinary numbers by vectors and matrices, you get a result where every element is multiplied by the number. To try this for yourself, type **a = [1,2;3,4] * 3** and press Enter. You see the following output:

```
a =
    3     6
    9    12
```

The example begins with the matrix, [1,2;3,4]. It then multiplies each element by 3 and places the result in a.

Division works in the same manner. To see how division works, type **b = [6, 9; 12, 15] / 3** and press Enter. You see the following output:

```
b =
    2     3
    4     5
```

Again, the example begins with a matrix, [6, 9; 12, 15], and right divides it by 3. The result is stored in b.

MATLAB supports both *right division,* where the left side is divided by the right side (what most people would consider the standard way of doing things), and *left division,* in which the left side is divided by the right side (also known as guzinta — goes into — division). When working with scalars, whether you use right division or left division doesn't matter. To see this fact for yourself, type **c = 3 \ [6, 9; 12, 15]** and press Enter. (Notice the use of the backslash, \, for left division.) You get the same result as before:

```
c =
    2     3
    4     5
```

Employing matrix multiplication

Multiplication occurs at several different levels in MATLAB. The following sections break down the act of matrix multiplication so that you can see each level in progression.

Multiplying two vectors

Vectors are just matrices of only one row or column. Remember that you create a row vector by separating values using a comma, such as [1, 2]. To create column vectors, you use a semicolon, such as [3; 4]. You can also use prime to create a row or column vector. For example, [3, 4]' is equivalent to [3; 4]. (Pay particular attention to the use of commas and semicolons.)

When you want to multiply one vector by another, you must have one row and one column vector. Try it for yourself by typing **d = [1, 2] * [3; 4]** and pressing Enter. You get the value 11 for output. Of course, the method used to perform the multiplication is to multiply the first element in the row vector by the first element of the column vector, and add the result to the multiplication of the second element of the row vector and the second element of the column vector. What you end up with is d = 1 * 3 + 2 * 4. This form of multiplication is also called an *inner product*.

It's also possible to create an *outer product* using MATLAB. In this case, each element in the first vector is multiplied by every element of the second vector (technically matrix multiplication), and the results of each multiplication are placed in a separate element. To put this in perspective, you'd end up with a 2 x 2 matrix consisting of [1 * 3, 2 * 3; 1 * 4, 2 * 4]. The easiest way to see how this works is by trying it yourself. Type **e = bsxfun(@times, [1, 2], [3; 4])** and press Enter. You see

```
e =
      3      6
      4      8
```

The bsxfun() function performs element-by-element operations. You supply a function name (or *handle*) to perform an element-by-element math operation on two objects (vectors in this case). We're using the @times function name, which performs multiplication. The two inputs are a row vector and a column vector. The output is a 2 x 2 matrix where the row 1 column 1 element is 1 * 3 (or the result of multiplying the first row element by the first column element). Likewise, the row 1 column 2 element is 2 * 3 (or the result of multiplying the second row element by the first column element). The second row multiplication works the same way as the first.

Another way to obtain the outer product is to ensure that the column vector appears first. For example, type **e = [3; 4] * [1, 2]** and you receive an output of

```
e =
      3      6
      4      8
```

Multiplying a matrix by a vector

When performing multiplication of a matrix by a vector, the order in which the vector appears is important. Row vectors appear before the matrix, but column vectors appear after the matrix. To see how the row vector approach works, type **f = [1, 2] * [3, 4; 5, 6]** and press Enter. You see an output of

```
f =
    13    16
```

The first element is produced by 1 * 3 + 2 * 5. The second element is produced by 1 * 4 + 2 * 6. However, the number of elements in the matrix must agree with the number of elements in the vector. For example, if the vector has three elements in a row, the matrix must have three elements in a column to match. To see how this works, type **g = [1, 2, 3] * [4, 5; 6, 7; 8, 9]** and press Enter. The result is

```
g =
    40    46
```

The number of elements in the output is controlled by the matrix in this case. For example, if the matrix were to have three elements in each row, the output would also have three elements. To see this principle in action, type **h = [1, 2, 3] * [4, 5, 6; 7, 8, 9; 10, 11, 12]** and press Enter. The result is

```
h =
    48    54    60
```

Working with a column vector is similar to working with a row vector, except that the position of the vector and matrix are exchanged. For example, if you type **i = [4, 5, 6; 7, 8, 9; 10, 11, 12] * [1; 2; 3]** and press Enter, you see this result:

```
i =
    32
    50
    68
```

Notice that the output is a column vector instead of a row vector. The result is produced by these three equations:

```
1 * 4 + 2 * 5 + 3 * 6
1 * 7 + 2 * 8 + 3 * 9
1 * 10 + 2 * 11 + 3 * 12
```

The order of the multiplication differs because you're using a column vector instead of a row vector. MATLAB produces the same result as you would get when performing the task using other means, but you need to understand how the data entry process affects the output.

Multiplying two matrices

When working with matrices, the number of rows in the first matrix must agree with the number of columns in the second matrix. For example, if the first matrix contains two rows containing three entries each, the second matrix must contain three rows and two entries each. To see this for yourself, type **j = [1, 2, 3; 4, 5, 6] * [7, 8; 9, 10; 11, 12]** and press Enter. You see the output as

```
j =
      58     64
     139    154
```

The output of the first column, first row is defined by 1 * 7 + 2 * 9, + 3 * 11. Likewise, the output of the second column, first row is defined by 1 * 8 + 2 * 10 + 3 * 12. The matrix math works just as you would expect.

Order is important when multiplying two matrices (just as it is when working with vectors). You can create the same two matrices, but obtain different results depending on order. If you reverse the order of the two matrices in the previous example by typing **k = [7, 8; 9, 10; 11, 12] * [1, 2, 3; 4, 5, 6]** and pressing Enter, you obtain an entirely different result:

```
k =
      39     54     69
      49     68     87
      59     82    105
```

Again, it pays to know how the output is produced. In this case, the output of the first column, first row is defined by 7 * 1 + 8 * 4. Likewise, the output of the second column of the first row is defined by 7 * 2 + 8 * 5.

Dividing two vectors

MATLAB will produce an output if you try to divide two vectors. For example, if you type **l = [2, 3, 4] / [5, 6, 7]** and press Enter, you receive a result of

```
l =
    0.5091
```

Likewise, you could try typing **l = [2, 3, 4] \ [5, 6, 7]** and press Enter. The results would be different:

```
l =
         0         0         0
         0         0         0
    1.2500    1.5000    1.7500
```

You get the same reproducible results every time, but you can see that they're interesting, at best. The output isn't useful because dividing two vectors isn't useful in general. You can read the details at http://van.physics.illinois.edu/qa/listing.php?id=24304 and many other places online. The point is that MATLAB tries to accommodate your needs, even if the result isn't particularly helpful.

Effecting matrix division

As with matrix multiplication, matrix division takes place at several different levels. The following sections explore division at each level.

Dividing a vector by a scalar

Dividing a vector by a scalar and producing a usable result is possible. For example, type **m = [2, 4, 6] / 2** and press Enter. You see the following result:

```
m =
    1    2    3
```

Each of the entries is divided by the scalar value. Notice that this is right division. Using left division (m = [2, 4, 6] \ 2) would produce an unusable result; however, using m = 2 \ [2, 4, 6] would produce the same result as before. MATLAB would do its best to accommodate you with a result, just not one you could really use. (See the "Dividing two vectors" sidebar for an explanation.)

Dividing a matrix by a vector

When dividing a matrix by a vector, defining the sort of result you want to see is important. Most people want to perform an element-by-element division. In this case, you use the bsxfun() function with the @rdivide function name — @rdivide for right division. To see how this works, type **n = bsxfun(@rdivide, [2, 4; 6, 8], [2, 4])** and press Enter. You see the following output:

```
n =
    1    1
    3    2
```

In this case, the element in column 1, row 1 is defined by 2 / 2. Likewise, the element in column 1, row 2 is defined by 6 / 2.

Dividing two matrices

When dividing two matrices, the dimensions of the two matrices must agree. For example, you can't divide a 3 x 2 matrix by a 2 x 3 matrix — both matrices must be the same dimensions, such as 3 x 2. To see how this works, type **o = [2, 4; 6, 8] / [1, 2; 3, 4]** and press Enter. You see the following result:

```
o =
    2    0
    0    2
```

Performing left division of two matrices is also possible. To see the result of performing left division using the same matrices, type **p = [2, 4; 6, 8] \ [1, 2; 3, 4]** and press Enter. Here's the result you see:

```
p =
    0.5000         0
         0    0.5000
```

It's essential to remember that matrix division isn't actually division as most people think of it. What you really do is multiply one matrix by the inverse of the other. For example, using the two matrices in this section, you can accomplish the same result of left division by typing **q = [2, 4; 6, 8] * inv([1, 2; 3, 4])** and pressing Enter. To perform right division, you simply change the inverted matrix by typing **r = inv([2, 4; 6, 8]) * [1, 2; 3, 4]** and pressing Enter. The inv() function always returns the inverse of the matrix that you provide as input, so you can use it to help you understand precisely how MATLAB is performing the task. However, using the inv() function is computationally inefficient. To make your scripts run faster, dividing is always better.

You can use the inv() function in many ways. For example, multiplying any matrix by its inverse, such as by typing **s = [1, 2; 3, 4] * inv([1, 2; 3, 4])**, yields the identity matrix.

What some people are actually looking for is element-by-element division. To accomplish this task, you must use the bsxfun() function. For example, to perform left division on the two preceding matrices, you type **t = bsxfun(@ldivide, [2, 4; 6, 8], [1, 2; 3, 4])** and press Enter. The result in this case is

```
t =
    0.5000    0.5000
    0.5000    0.5000
```

Likewise, you can perform right division. To see how this works, type **u = bsxfun(@rdivide, [2, 4; 6, 8], [1, 2; 3, 4])** and press Enter. You see the following output:

```
u =
     2     2
     2     2
```

Creating powers of matrices

Sometimes you need to obtain the power or root of a matrix. MATLAB provides several different methods for accomplishing this task. The most common method is to use the circumflex (^) to separate the matrix from the power to which you want to raise it. To see how this works, type **v = [1, 2; 3, 4]^2** and press Enter. The output is the original matrix squared, as shown here:

```
v =
     7    10
    15    22
```

You can obtain the same result using the mpower() function. Try it by typing **w = mpower([1, 2; 3, 4], 2)** and pressing Enter. You see the same output as when using the circumflex.

To obtain the root of a matrix, you use a fractional value as input. For example, to obtain the square root of the previous example, you use a value of 0.5. To see this feature in action, type **x = [1, 2; 3, 4]^0.5** and press Enter. You see the following output:

```
x =
    0.5537 + 0.4644i    0.8070 - 0.2124i
    1.2104 - 0.3186i    1.7641 + 0.1458i
```

It's even possible to obtain the inverse of a matrix by using a negative power. For example, try typing **z = [1, 2; 3, 4]^(–1)** and pressing Enter (notice that the –1 is enclosed in parenthesis to avoid confusion). You see the following output:

```
z =
   -2.0000     1.0000
    1.5000    -0.5000
```

MATLAB also provides the means for performing an element-by-element power or root of a matrix using the bsxfun() function and the @power handle. To see how this works, type **aa = bsxfun(@power, [1, 2; 3, 4], 2)** and press Enter. You see the following output, in which each element is multiplied by itself:

```
aa =
     1     4
     9    16
```

Working element by element

A number of previous sections describe how to use the bsxfun() function to perform tasks element by element. For example, to find the square of the matrix [1, 2; 3, 4], you type **aa = bsxfun(@power, [1, 2; 3, 4], 2)** and press Enter. Of course, the bsxfun() function provides all sorts of function handles, and you can see them all by typing **help('bsxfun')** and pressing Enter.

The problem is that the bsxfun() function requires quite a bit of typing, so you might not want to use it all the time. An alternative to using this function involves using the dot (.) operator. For example, to obtain the square of the previous matrix using the dot operator, you type **ab = [1, 2; 3, 4].^2** and press Enter. The output is as you expect:

```
ab =
     1     4
     9    16
```

Checking matrix relations

This chapter discusses a number of techniques to perform any given task. For example, you can create the inverse of a matrix using the `inv()` function, or you can simply set it to a power of –1. The problem is that you don't really know that they are equal outputs. The `bsxfun()` comes in handy for all sorts of tasks, and checking for equality is yet another way you can use it. To see for yourself that `inv()` and a power of –1 produce the same result, simply type **bsxfun(@eq, inv([1, 2; 3, 4]), [1, 2; 3, 4]^(-1))** and press Enter. The output you see is

```
ans =
     1     1
     1     1
```

The `@eq` function handle tells `bsxfun()` to check for equality. Each element is compared. When the elements compare, the output is 1. So, a matrix output of 1s tells you that all of the elements compared in this case. You can perform other relational checks using `bsxfun()` with the following function handles:

- ✔ `@eq`: Equal
- ✔ `@ne`: Not equal
- ✔ `@lt`: Less than
- ✔ `@le`: Less than or equal
- ✔ `@gt`: Greater than
- ✔ `@ge`: Greater than or equal

Notice that the dot goes between the matrix and the circumflex. You can use the dot operator in every other circumstance you can think of to modify MATLAB behavior to work element by element. For example, to perform element-by-element multiplication, you place the dot operator in front of the multiplication operator. To try the multiplication, type **ac = [1, 2; 3, 4] .* [5, 6; 7, 8]** and press Enter. You see the following output:

```
ac =
     5    12
    21    32
```

The dot operator always precedes the task operator that you want to use. Even if there is a space between the matrix and the task operator, the dot operator must appear with the task operator without a space, such as .* for multiplication.

Using complex numbers

Complex numbers consist of a real part and an imaginary part (see http:// www.mathsisfun.com/numbers/imaginary-numbers.html for a quick overview of imaginary numbers). MATLAB uses the i and j constants to specify the imaginary part of the number. For example, when you compute the square root of the matrix [1, 2; 3, 4], you obtain an output that contains imaginary numbers. To see this for yourself, type **ad = [1, 2; 3, 4]^0.5** and press Enter. You see the following result:

```
ad =
   0.5537 + 0.4644i   0.8070 - 0.2124i
   1.2104 - 0.3186i   1.7641 + 0.1458i
```

The first column of the first row contains a real value of 0.5537 and an imaginary value of 0.4644i. The i that appears after the value 0.4644 tells you that this is an imaginary number. The j constant means the same thing as the i constant, except that the j constant is used in electronics work (i is already used to represent current).

You can perform tasks with imaginary numbers just as you would any other number. For example, you can square the ad matrix by typing **ae = ad^2** and pressing Enter. The result might not be what you actually wanted, though:

```
ae =
   1.0000 + 0.0000i   2.0000 + 0.0000i
   3.0000 - 0.0000i   4.0000 + 0.0000i
```

After a matrix includes imaginary numbers, you need to convert them to obtain a desired format. For example, if you type **af = int32(ad^2)** and press Enter, you obtain the desired result, shown here:

```
af =
           1           2
           3           4
```

The int32() function performs the required conversion process for you. Of course, using int32(), or any other function of the same type, at the wrong time can result in data loss. For example, if you type **ag = int32([1, 2; 3, 4]^0.5)** and press Enter, you lose not only the imaginary part of the number but the fractional part as well. The output looks like this:

```
ag =
           1           1
           1           2
```

MATLAB assumes that you know what you're doing, so it doesn't stop you from making critical errors. The output conversion functions are

✔ double()

✔ single()

✔ int8()

✔ int16()

✔ int32()

✔ int64()

✔ uint8()

✔ uint16()

✔ uint32()

✔ uint64()

Working with exponents

You use matrix exponential to perform tasks such as solving differential equations (read about them at `http://www.sosmath.com/matrix/expo/expo.html`). MATLAB provides two functions for working with exponents. The first is the `expm()` function, which performs a standard matrix exponential. For example, when you type **ah = expm([1, 2; 3, 4])** and press Enter, you see this result:

```
ah =
    51.9690    74.7366
   112.1048   164.0738
```

MATLAB also makes it easy to perform element-by-element exponential using the `exp()` function. To see how this works, type **ai = exp([1, 2; 3, 4])** and press Enter. You see the following output:

```
ai =
     2.7183     7.3891
    20.0855    54.5982
```

Working with Higher Dimensions

A vector is one dimensional — just one row or one column. A matrix is a two-dimensional table, much like the kind you're used to with Excel spreadsheets, with rows being one dimension and columns being the second. You can go as high as you want. If a matrix is like a page in a book (a table — two dimensions), three dimensions is like the book itself, and four like a shelf of books. In fact, there is no limit to the number of dimensions you can use to express an idea or data element.

Images are an example of computational objects that rely on more than one dimension:

✔ The first dimension is the x coordinate of a pixel.

✔ The second dimension is the y coordinate of a pixel.

✔ The third dimension is the pixel color.

Now that you have a better idea of how you might use more than just two dimensions, it's time to see how you can implement them. The following sections describe how to work with multiple dimensions when using MATLAB.

Creating a multidimensional matrix

MATLAB provides a number of ways in which to create multidimensional arrays. The first method is to simply tell MATLAB to create it for you and fill each of the elements with zeros. The zeros() function helps you perform this task. To create a 2 x 3 x 3 matrix, you type aj = zeros(2, 3, 3) and press Enter. You see the following output:

```
aj(:,:,1) =
     0      0      0
     0      0      0
aj(:,:,2) =
     0      0      0
     0      0      0
aj(:,:,3) =
     0      0      0
     0      0      0
```

This output tells you that there are three stacked 2 x 3 matrices and each one is filled with zeros. Of course, you might not want to start out with a matrix that's filled with zeros, so you can use another approach. The following steps help you create a 2 x 3 x 3 matrix that is already filled with data:

1. **Type** ak(:,:,1) = [1, 2, 3; 4, 5, 6] **and press Enter.**

   You see the following result:

   ```
   ak =
        1      2      3
        4      5      6
   ```

   This step creates the first page of the three dimensional matrix. You want three pages, so you actually need to perform this step three times.

2. **Type** ak(:,:,2) = [7, 8, 9; 10, 11, 12] **and press Enter.**

MATLAB adds another page, as shown:

```
ak(:,:,1) =
       1        2        3
       4        5        6
ak(:,:,2) =
       7        8        9
      10       11       12
```

If you look at the Workspace window at this point, you see that the size column for ak is now 2 x 3 x 2. It's at this point that you see the third dimension added. Before you added this second page, MATLAB simply treated ak as a 2 x 3 matrix, but now it has the third dimension set.

3. Type ak(:,:,3) = [13, 14, 15; 16, 17, 18] **and press Enter.**

The output now looks much like the aj output, except that the elements have values, as shown here:

```
ak(:,:,1) =
       1        2        3
       4        5        6
ak(:,:,2) =
       7        8        9
      10       11       12
ak(:,:,3) =
      13       14       15
      16       17       18
```

You don't have to define assigned values using multiple steps. The cat() function lets you create the entire three-dimensional matrix in one step. The first entry that you make for the cat() function is the number of dimensions. You then add the data for each dimension, separated by commas. To see how this works, type **al = cat(3, [1, 2, 3; 4, 5, 6], [7, 8, 9; 10, 11, 12], [13, 14, 15; 16, 17, 18])** and press Enter. You see this output (which looks amazingly like the ak matrix):

```
al(:,:,1) =
       1        2        3
       4        5        6
al(:,:,2) =
       7        8        9
      10       11       12
al(:,:,3) =
      13       14       15
      16       17       18
```

You may also decide that you don't want to type that much but still don't want zeros in the matrix. In this case, use the randn() function for random normally distributed data or the rand() function for uniformly distributed data. This function works just like the zeros() function, but it fills the elements with random data.

To see how this function works, type **am = randn(2, 3, 3)** and press Enter. You see a three-dimensional array filled with random data. It's not likely that your output will look precisely like the following output, but the following output does provide an idea of what you should expect:

```
am(:,:,1) =
    1.4090    0.6715    0.7172
    1.4172   -1.2075    1.6302
am(:,:,2) =
    0.4889    0.7269    0.2939
    1.0347   -0.3034   -0.7873
am(:,:,3) =
    0.8884   -1.0689   -2.9443
   -1.1471   -0.8095    1.4384
```

Accessing a multidimensional matrix

No matter how you create the matrix, eventually you need to access it. To access the entire matrix, you simply use the matrix name, as usual. However, you might not need to access the entire matrix. For example, you might need to access just one page. The examples in this section assume that you created matrix ak in the previous section. To see just the second page of matrix ak, you type **ak(:, :, 2)** and press Enter. Not surprisingly, you see the second page, as shown here:

```
ans =
     7     8     9
    10    11    12
```

The colon (:) provides a means for you to tell MATLAB that you want the entire range of a matrix element. The values are rows, columns, and pages in this case. So the request you made was for the entire range of page 2. You could ask for just a row or column. To get the second row of page 2, you type **ak(2, :, 2)** and press Enter. The output looks like this:

```
ans =
    10    11    12
```

The second column of page 2 is just as easy. In this case, you type **ak(:, 2, 2)** and press Enter. The output appears in column format, like this:

```
ans =
     8
    11
```

Accessing an individual value means providing all three values. When you type **ak(2, 2, 2)** and press Enter, you get 11 as the output because that's the value in row 2, column 2, of page 2 for matrix ak.

You also have access to range selections for multidimensional matrices. In this case, you must provide a range for one of the entries. For example, if you want to obtain access to row 2, columns 1 and 2, of page 2 for matrix ak, you type **ak(2, [1:2], 2)** and press Enter. Notice that the range appears within square brackets, and the start and end of the range are separated by a colon. Here is the output you see in this case:

```
ans =
    10    11
```

The use of ranges works wherever you need them. For example, say that you want rows 1 and 2, columns 1 and 2, of page 2. You type **ak([1:2], [1:2], 2)** and press Enter. The result looks like this:

```
ans =
     7     8
    10    11
```

Replacing individual elements

As you work through problems and solve difficulties, you might find changing some of the data in a matrix necessary. The problem is that you don't want to have to re-create the matrix from scratch just to replace one value. Fortunately, you can replace individual values in MATLAB. The examples in this section assume that you created matrix ak in the "Creating a multidimensional matrix" section, earlier in this chapter.

The previous section tells you how to access matrix elements. You use this ability to change values. For example, the value in row 2, column 2, of page 2 in matrix ak is currently set to 11. You may decide that you really don't like the number 11 there and want to change it to 44 instead. To perform this task, type **ak(2, 2, 2) = 44** and press Enter. You see the following result:

```
ak(:,:,1)  =
     1     2     3
     4     5     6
ak(:,:,2)  =
     7     8     9
    10    44    12
ak(:,:,3)  =
    13    14    15
    16    17    18
```

Notice that MATLAB displays the entire matrix. Of course, you may not want to see the entire matrix every time you replace a single value. In this case, end the command with a semicolon. When you type **ak(2, 2, 2) = 44;** and press Enter, the change still takes place, but you don't see the result onscreen. For now, continuing to display the information is a good idea so that you can tell whether you have entered the commands correctly and have obtained the desired result.

Replacing a range of elements

If you have a number of values to replace in a matrix, replacing them one at a time would become boring. More important, you start to make mistakes after a while and your results don't come out as you thought they would. Replacing a range of values with a single command is the best idea in this case. The examples in this section assume that you created matrix ak in the "Creating a multidimensional matrix" section, earlier in this chapter.

You have many different ways to make replacements to a range of elements in your existing matrix. Of course, before you can replace a range of elements, you need to know how to access them. The "Accessing a multidimensional matrix" section, earlier in this chapter, tells you how to access matrix elements.

You can make a single value replacement for a range. Say that you want to replace row 2, columns 1 and 2, of page 2 with the number 5. To perform this task, type **ak(2, [1:2], 2) = 5** and press Enter. The single value appears in both places, as shown in this output:

```
ak(:,:,1) =
       1        2        3
       4        5        6
ak(:,:,2) =
       7        8        9
       5        5       12
ak(:,:,3) =
      13       14       15
      16       17       18
```

Of course, a single value replacement might not work. You can also create range replacements in which you replace each element with a different value. For example, you might want to replace row 2, column 1, of page 2 with the number 22, and row 2, column 2, of page 2 with the number 33. To perform this task, you type **ak(2, [1:2], 2) = [22, 33]** and press Enter. Here is the output you see:

```
ak(:,:,1) =
       1        2        3
       4        5        6
ak(:,:,2) =
       7        8        9
      22       33       12
ak(:,:,3) =
      13       14       15
      16       17       18
```

Column changes work the same way. In this case, you might want to replace row 1, column 3, of page 2 with the number 44, and row 2, column 3, of page 2 with the number 55. To perform this task, you type **ak([1:2], 3, 2) = [44, 55]** and press Enter. Notice that you didn't have to define the input vector using a column format. Here's the result you see:

```
ak(:,:,1) =
     1       2       3
     4       5       6
ak(:,:,2) =
     7       8      44
    22      33      55
ak(:,:,3) =
    13      14      15
    16      17      18
```

When replacing a rectangular range, you need to use a proper matrix for input. For example, you might want to replace a rectangular range between columns 1 and 2, rows 1 and 2, of page 1 with the values 11, 22, 33, and 44. To perform this task, you type **ak([1:2], [1:2], 1) = [11, 22; 33, 44]** and press Enter. Here's the result you see:

```
ak(:,:,1) =
    11      22       3
    33      44       6
ak(:,:,2) =
     7       8      44
    22      33      55
ak(:,:,3) =
    13      14      15
    16      17      18
```

Modifying the matrix size

You might not think that resizing a matrix is possible, but MATLAB can do that, too. It can make the matrix larger or smaller. The technique for making the matrix smaller is a bit of a trick, but it works well, and you likely will have a need for it at some point. The examples in this section assume that you created matrix ak in the "Creating a multidimensional matrix" section, earlier in this chapter.

As with range replacement, you need to know how to access ranges before you start this section. The "Accessing a multidimensional matrix" section, earlier in this chapter, tells you how to access matrix elements.

The current ak matrix is 2 x 3 x 3. You might want to add another row, even if that row consists only of zeros, to make the matrix square for some advanced task you need to perform. Some tasks work properly only with square matrices, so this is a real concern. To add another row to the existing matrix, type **ak(3, :, :) = 0** and press Enter. You see the following result:

```
ak(:,:,1) =
      11        22         3
      33        44         6
       0         0         0
ak(:,:,2) =
       7         8        44
      22        33        55
       0         0         0
ak(:,:,3) =
      13        14        15
      16        17        18
       0         0         0
```

All three pages now have another row. However, you might decide that you really don't want that extra row after all. To delete the row, you need to perform a bit of a trick — you set the row to a null (empty) value using an empty matrix ([]). To see how this works, type **ak(3, :, :) = []** and press Enter. You see the following result:

```
ak(:,:,1) =
      11        22         3
      33        44         6
ak(:,:,2) =
       7         8        44
      22        33        55
ak(:,:,3) =
      13        14        15
      16        17        18
```

At this point, you probably wonder what would happen if you added a column or row to just a single page. Try typing **ak(:, 4, 1) = [88, 99]** and pressing Enter. This command adds a fourth column to just page 1 and fills it with the values 88 and 99. MATLAB provides the following output:

```
ak(:,:,1) =
      11        22         3        88
      33        44         6        99
ak(:,:,2) =
       7         8        44         0
      22        33        55         0
ak(:,:,3) =
      13        14        15         0
      16        17        18         0
```

Notice that the other pages also have a fourth column now. The column is filled with zeros, but MATLAB automatically adds it for you to keep things tidy.

Using cell arrays and structures

The matrices you have created so far all contain the same data type, such as `double` or `uint8`. Every matrix you create will contain data of the same type — you can't mix types in a matrix. You do have, however, two other means to store data:

- A *cell array* works much like a spreadsheet
- A *structure* works much like a database record

These two containers let you store other kinds of data, and mix and match types as needed. Theoretically, you could use them to create a small database or some sort of alternative storage on your machine without resorting to another application. However, if you're a typical user, you probably won't use these structures, but at least knowing what they are is a good idea. The following sections provide an introduction and point you to more help in case you need to know more.

Understanding cell arrays

Cell arrays are naturals for spreadsheets because an individual cell in a cell array is like a cell in a spreadsheet. In fact, when you import a spreadsheet into MATLAB, each cell in the spreadsheet becomes a cell in a MATLAB cell array. Because spreadsheets are so popular, you're more likely to encounter a cell array than a structure.

You use the `cell()` function to create a new cell array. For example, to create a 2 x 2 x 2 cell array, you type **an = cell(2, 2, 2)** and press Enter. You see this result:

```
an(:,:,1) =
      []      []
      []      []
an(:,:,2) =
      []      []
      []      []
```

The cells are empty at this point. Cell arrays rely on a different kind of bracket to provide access to individual elements, the curly braces ({}). In order to make the an cell array useful, begin by typing the following lines of code, pressing Enter after each line:

```
an{1,1,1}='George';
an{1,2,1}='Smith';
an{2,1,1}=rand();
an{2,2,1}=uint16(1953);
an{1,1,2}=true;
an{1,2,2}=false;
an{2,1,2}=14.551+2.113i;
an{2,2,2}='The End!'
```

Because all the lines except for the last one ended with a semicolon, you didn't see any output. However, after you type the last line, you see the following output from MATLAB:

```
an(:,:,1) =
    'George'      'Smith'
    [0.6948]      [ 1953]
an(:,:,2) =
    [                1]    [          0]
    [14.5510 + 2.1130i]    'The End!'
```

The output looks just like any other multidimensional matrix. You can access it the same way, except that you use curly braces. For example, type **an{1, :, 2}** and press Enter to see the first row of page 2. The result looks like this:

```
ans =
     1
ans =
     0
```

MATLAB uses the values 1 and 0 to represent `true` and `false`. To test this fact for yourself, type **true** and press Enter. You see an output value of 1. Likewise, type **false** and press Enter. You see an output value of 0.

Each of the entries is treated as a separate item, but you can select ranges and work with individual values, just as you do when working with a multidimensional matrix. However, you must use the curly braces when working with cell arrays.

You can distinguish between cell arrays and matrices in the Workspace window by the icons they use. The cell array icon contains a pair of curly braces, so it contrasts well with the matrix icon, which looks like a little mini table. The Value column also specifically tells you that the entry is a cell rather than a specific data type, such as a double.

Understanding structures

Structures are more closely related to SQL database tables than spreadsheets. Each entry consists of a field name and value pair. The field names are generally descriptive strings, but the values can be anything that relates to that field name. To get a better idea of how a structure works, type **MyStruct = struct('FirstName', 'Amy', 'LastName', 'Jones', 'Age', 32, 'Married', false)** and press Enter. You see the following output:

```
MyStruct =
    FirstName: 'Amy'
     LastName: 'Jones'
          Age: 32
      Married: 0
```

Notice how the field names are paired with their respective values. A structure is designed to reside in memory like a database. Currently, MyStruct has just one record in it. You can access this record by typing **MyStruct(1)** and pressing Enter. The results are as follows:

```
ans =
     FirstName: 'Amy'
      LastName: 'Jones'
           Age: 32
       Married: 0
```

Dealing with an entire record probably isn't what you had in mind, though. To access a particular field, you type a period, followed by the field name. For example, type **MyStruct(1).LastName** and press Enter to access the LastName field. You get the following answer:

```
ans =
Jones
```

A single record structure isn't very useful. You might have quite a few records in a real structure. To add another record to MyStruct, type **MyStruct(2) = struct('FirstName', 'Harry', 'LastName', 'Smith', 'Age', 35, 'Married', true)** and press Enter. The output might surprise you this time. You see

```
MyStruct =
1x2 struct array with fields:
     FirstName
     LastName
     Age
     Married
```

The output tells you how many records are in place. You can test for the second record by typing **MyStruct(2)** and pressing Enter. The output is precisely as you expect:

```
ans =
     FirstName: 'Harry'
      LastName: 'Smith'
           Age: 35
       Married: 1
```

Don't limit your input of structures to the common data types. Structure data may contain a matrix, even multidimensional matrices, and you can mix sizes. In addition, structures and cell arrays can contain each other. An element in a structure can be a cell array, and a cell in a cell array can be a structure. The point is that these are extremely flexible ways to store information when you need them; however, you shouldn't make things overly complex by using them when you don't need them. If you can create storage that uses one common data type, matrices are the way to go.

This is only a brief overview of structures. Go to MATLAB's help system and click Matlab⇨Language Fundamentals⇨Data Types⇨Structures to find additional information on this topic.

Using the Matrix Helps

As you work with matrices, you may need to test your code, and MATLAB has provided some help in the form of ways to create a matrix (Table 5-1), test matrices (Table 5-2), and diagnose matrix problems (Table 5-3). The tables in this section help you work more productively with matrices and get them working considerably faster.

The tables contain only the more useful commands. MATLAB has a lot more to offer. The following locations in MATLAB's help system can provide you with substantially more information:

✔ Help Home⇨MATLAB⇨Language Fundamentals⇨Matrices and Arrays

✔ Help Home⇨MATLAB⇨Mathematics⇨Elementary Math⇨Constants and Text Matrices

Table 5-1		Matrix Creation	
Function	*What It Does*	*Generic Call*	*Example*

zeros()	Creates a matrix of all zeros	zeros(<mat_size>), where <mat_size> is a positive integer number, two number arguments, or a vector of numbers.	>> zeros(3)
			ans =
			  0 0 0
			  0 0 0
			  0 0 0

ones()	Creates a matrix of ones	ones(<mat_size>), where <mat_size> is a positive integer number, two number arguments, or a vector of numbers.	>> ones(3)
			ans =
			  1 1 1
			  1 1 1
			  1 1 1

Function	What It Does	Generic Call	Example
eye()	Creates an identity matrix with one on the main diagonal and zero elsewhere	eye(<mat_size>), where <mat_size> is a positive integer number, two number arguments, or a vector of numbers. This call doesn't allow you to create N-dimensional arrays.	>>eye(3) ans= 1 0 0 0 1 0 0 0 1
rand()	Creates a matrix of uniformly distributed random numbers	rand(<mat_size>), where <mat_size> works like the argument(s) of eye.	>>rand(3) ans= 0.8147 0.9134 0.2785 0.9058 0.6324 0.5469 0.1270 0.0975 0.9575
randn()	Creates a matrix of normally distributed random numbers (mean=0, SD=1)	randn(<mat_size>), where <mat_size> works like the argument(s) of eye.	>> randn(3) ans = 0.5377 0.8622 -0.4336 1.8339 0.3188 0.3426 -2.2588 -1.3077 3.5784
blkdiag()	Makes a block diagonal matrix	blkdiag(a, b, c, ...), where a, b, c, ... are matrices.	>> blkdiag(ones(2), ones(2)) ans = 1 1 0 0 1 1 0 0 0 0 1 1 0 0 1 1

Table 5-2		**Test Matrices**	
Function	What It Does	Generic Call	Example
magic()	Creates a magic square matrix — the sum of rows and columns are equal	magic(n), where n is the number of rows and columns.	>> magic(3) ans = 8 1 6 3 5 7 4 9 2

(continued)

Table 5-2 *(continued)*

Function	What It Does	Generic Call	Example
| gallery() | Produces a wide variety of test matrices for diagnosis of your code | Gallery(... '<option>',... <mat_size>, j), where 'option' is a string that defines what task to perform, such as binomial, which creates a binomial matrix. <mat_size> is a positive integer number, two number arguments, or a vector of numbers. Each different positive integer *j* produces a different matrix. | ```
>> gallery('normal
data',3,3)
ans =
 0.9280 -0.7230 0.2673
 0.1733 -0.5744 1.3345
-0.6916 -0.3077 -1.3311
``` |

Table 5-3   **Helpful Commands**

Function	What It does	Generic Call	Example
rng()	Controls the random number generator	rng(<my_seed>, '<my_option>'), where <my_seed> is a numeric value used to define the starting point for random values and '<my_option>' is the option used to set the random number generator.	rng('default') resets the random number generator to a known value. This command is useful to reproduce random matrices.
| size() | Returns the size of a matrix | size(<your_ matrix>) | ```
>> size(zeros([2,3,4]))
ans =
 2 3 4
``` |

Function	What It does	Generic Call	Example
length()	Returns the length of a vector	length(<*your_ matrix*>)	>> length(0:50) ans = 51
spy()	Produces a figure identifying where zeros are in a matrix	spy(<*your_ matrix*>)	>> spy(blkdiag (ones(100),... ones(200),ones(100)))

Chapter 6

Understanding Plotting Basics

MATLAB includes fabulous routines for plotting (or graphing) the data and expressions that you supply to the software. Using MATLAB's familiar interface, you can produce visual representations of various functions and data sets, including 2D x-y graphs, log scales, bar, and polar plots, as well as many other options. The visuals that MATLAB produces resemble anything from the graph of an algebraic equation to pie charts often used in business and to specialized graphs.

In this chapter, you find out how to use 2D plotting functions to create expression and data plots and how the same process works with other plotting routines in MATLAB. You also discover the commonly used visual styles for representing various types of data, how to combine plots, and how to modify the plots to match specific data sets.

With MATLAB, you can create plots based purely on the formula you provide. Although this chapter focuses on the more commonly used vector and matrix inputs, Appendix B provides a listing of all the plot types that MATLAB supports. Be sure to also check out the online materials for this book (as described in this book's Introduction) and the blog posts at `http://blog.johnmuellerbooks.com` to see how to work with other plot types.

Considering Plots

A *plot* is simply a visualization of data. Most people see a series of numbers in a table and can't really understand their meaning. Interpreting what the data means is hard to do without thinking about the relationship between data points. A plot makes the relationships between data points more obvious to the viewer and helps the viewer see patterns in the data. The following sections help you discover how MATLAB plots are special and can make the visualization of your data interesting and useful.

Understanding what you can do with plots

People are visually oriented. You could create a standard table showing the data points for a sine wave and have no one really understand that it was a sine *wave* at all or that the data points *move* in a certain way. However, if you plot that information, it becomes apparent to everyone that a sine wave has a particular presentation and appearance. The pattern of the sine wave becomes visible and understandable.

A sine wave consists of a particularly well-known set of data points, so some people might recognize the data for what it is. As your data becomes more complex, however, recognizing the patterns becomes more difficult — to the point at which most people won't understand what they're seeing. So the first goal of a plot is to make the pattern of data readily apparent.

Presentation is another aspect of plotting. You can take the same data and provide multiple views of it to make specific points — the company hasn't lost much money on bad widgets, for example, or the company has gained quite a few new customers due to some interesting research. Creating the right plot for your data defines a specific view of the data: It helps you make your point about whatever the data is supposed to represent.

Creative interaction with the data is another reason to use plots. People see not only the patterns that are present in plots but also see the ones that could be present given the right change in conditions. It's the creative interaction that makes plotting data essential for scientists and engineers. The ability to see beyond the data is an important part of the plotting process.

Comparing MATLAB plots to spreadsheet graphs

Although it might seem obvious at first, spreadsheet graphs are generally designed for use in business. As a result, the tools you find are better suited to making a point about some business need, such as this quarter's sales or the project production rate in the factory. A spreadsheet graph includes the tools of business, such as the need to add trend lines of various sorts to show how the numbers are changing over time.

MATLAB plots are more suited to scientific and engineering needs. A MATLAB plot does include some of the same features as a spreadsheet graph. For example, you can create a pie chart in either environment and assign data points to the chart in about the same manner. However, MATLAB includes plots that you can't find in the business environment, such as a semilogx (used to plot logarithmic data). A business user probably wouldn't have much need for a stem plot — the plot that shows the frequency at which certain values appear.

The way in which the two environments present information differs as well. A spreadsheet graph is designed to present an overview in an aesthetically pleasing manner. The idea is to convince a viewer of the validity of the data by showing general trends. Business users tend not to have time to dig into the details; they need to make decisions quickly based on trends. MATLAB graphs are all about the details. With this in mind, you can zoom in on a graph, examine individual data points, and work the plot in ways that a business user doesn't require.

No best approach to presenting information in graphic form exists. The only thing that matters is displaying the information in a manner that most helps the viewer. The essential difference in the two environments is that one allows the viewer to make decisions quickly and the other allows the viewer to make decisions accurately. Each environment serves its particular user's needs.

Creating a plot using commands

MATLAB makes creating a plot easy. Of course, before you can create any plot, you need a source of data to plot. The following steps help you create a data source and then use that data source to generate a plot. Even though MATLAB's plotting procedure looks like a really simplistic approach, it's actually quite useful for any data you want to plot quickly. In addition, it demonstrates that you don't even have to open any of the plotting tools to generate a plot in MATLAB.

1. **Type** x = -pi:0.01:pi; **and press Enter in the Command window.**

   MATLAB generates a vector, x, and fills it with a range of data points for you. The data points begin at –pi and end at pi, using 0.01 steps. The use of the semicolon prevents the output of the data points to the Command window, but if you look in the Workspace window, you see that the vector has 629 data points.

2. **Type** plot(x, sin(x)), grid on **and press Enter.**

You see the plot shown in Figure 6-1 appear. It's a sine wave created by MATLAB using the input you provided.

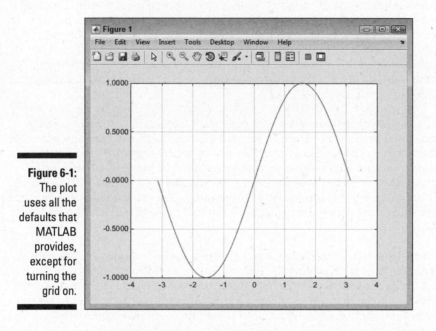

Figure 6-1:
The plot uses all the defaults that MATLAB provides, except for turning the grid on.

The plot() function accepts the data point entries that you provide. The vector x contains a series of values between –pi and pi. Taking the sine of each of these values using the sin() function creates the values needed to generate the plot shown. This version of the plot() function shows the minimum information that you can provide. The x value that appears first contains the information for the x-axis of the plot. The sin(x) entry that appears second contains the information for the y-axis of the plot.

It's possible to create any sort of plot using commands just as it is to use the graphic aids (such as the GUI shown in Figure 6-1) that MATLAB provides. For example, type **area(x,sin(x)), grid** and press Enter. You see the plot shown in Figure 6-2. However, this time the sine wave is shown as an area plot. MATLAB also has methods for modifying the appearance of the plot using commands.

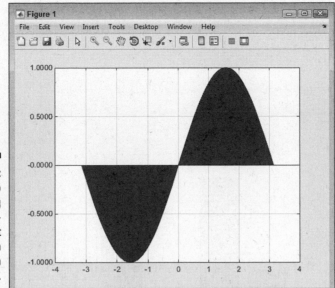

Figure 1
File   Edit   View   Insert   Tools   Desktop   Window   Help

Figure 6-2:
You can do
anything
with com-
mands that
you can
do with
the GUI.

Creating a plot using the Workspace window

The Workspace window displays all the variables that you create, no matter what type they might be. What you may not realize is that you can right-click any of these variables and create a plot from them. (If you don't see your plot listed, select the Plot Catalog option to see a full listing of the available plots.) The following steps help you create a variable and then plot it using the Workspace window functionality.

1. **Type** y = [5, 10, 22, 6, 17]; **and press Enter in the Command window.**

    You see the variable y appear in the Workspace window.

2. **Right-click** y **in the Workspace window and choose** bar(y) **from the context menu that appears.**

    MATLAB creates a bar graph using the default settings, as shown in Figure 6-3.

Even though this method might seem really limited, it's a great way to create a quick visualization of data so that you can see patterns or understand how the various data points interact. The advantage of this method is that it's quite fast.

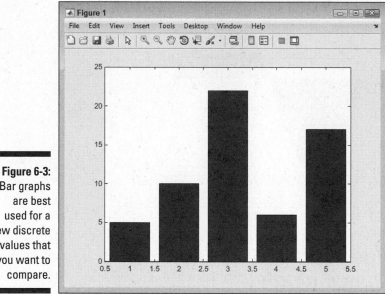

Figure 6-3:
Bar graphs
are best
used for a
few discrete
values that
you want to
compare.

MATLAB overwrites the previous plot you create when you create a new plot unless you use the `hold` command that is described later in the chapter. If you created the examples in the previous section, you should note that all the plots have appeared in the Figure 6-1 window and that no new plot windows have been created. Your old plot is immediately overwritten when you create a new one unless you save the old plot to disk or use the `hold` command.

Creating a plot using the Plots tab options

When you view the Plots tab in MATLAB, you see a gallery of the kinds of plots you can create. You initially see just a few of the available plots. However, if you click the downward-pointing arrow button at the right side of the gallery, you see a selection of plot types like the one shown in Figure 6-4.

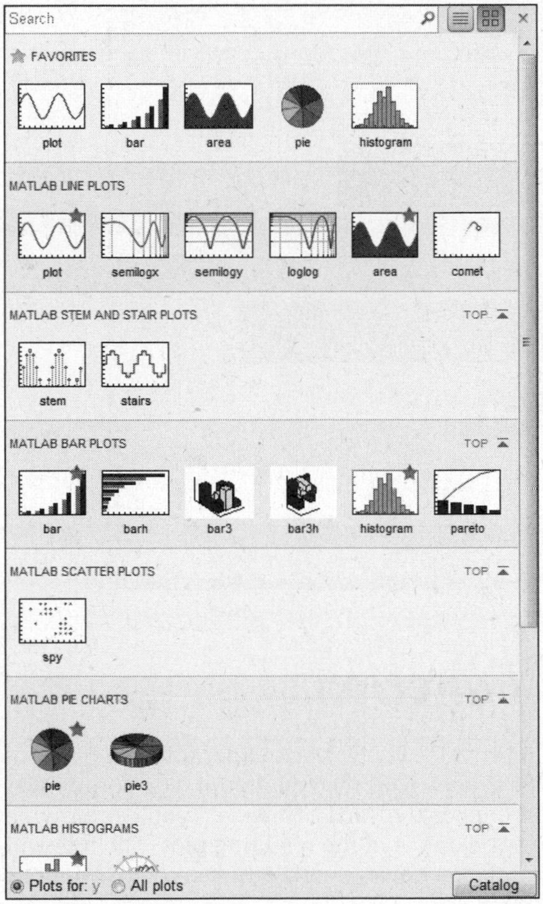

Figure 6-4:
MATLAB comes with a large number of plot types that you can use.

To use this feature, select a variable in the Workspace window and then choose one of the plots from the gallery list. This is the technique to use if you can't quite remember what sort of plot you want to create (making the command option less convenient) and the option doesn't appear in the Workspace window context menu. For example, you might want to create a horizontal bar plot using variable y. To perform this task, simply click variable y in the Workspace window and then choose barh in the MATLAB Bar Plots section of the gallery. The output MATLAB comes up with looks like Figure 6-5.

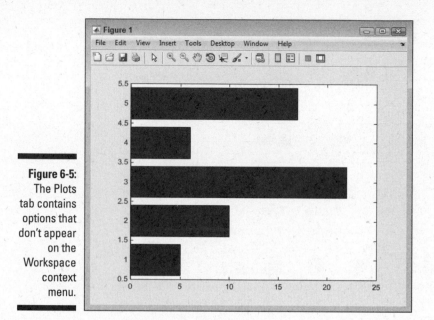

Figure 6-5:
The Plots
tab contains
options that
don't appear
on the
Workspace
context
menu.

Using the Plot Function

The plot() function provides you with considerable flexibility in using commands to create and modify a plot. As a minimum, you supply two vectors: one for the x axis and one for the y axis. However, you can provide more information to adjust the appearance of the resulting plot. The following sections provide additional details on how to work with the plot() function and make it provide the output you want.

Working with line color, markers, and line style

The plot() function can actually accommodate values in groups of three: the x axis, the y axis, and a character string that specifies the line color, marker style, and line style. Table 6-1 shows the values for the character string; you'd use values from each of the three entries (x axis, y axis, character string) to change the appearance of the plot.

Table 6-1		Line Color, Data Point Style, and Line Style			
Color		**Marker**		**Style**	
Code	**Line Color**	**Code**	**Marker Style**	**Code**	**Line Style**
b	blue	.	point	-	Solid
g	green	o	circle	:	Dotted
r	red	x	x-mark	-.	dash dot
c	cyan	+	plus	--	Dashed
m	magenta	*	star	(none)	no line
y	yellow	s	square		
k	black	d	diamond		
w	white	v	down triangle		
		^	up triangle		
		<	left triangle		
		>	right triangle		
		p	5 point star		
		h	6 point star		

You can combine the entries in various ways. For example, type **plot(1:5, y, 'r+–')** and press Enter to obtain the plot shown in Figure 6-6. Even though you can't see it in the book, the line is red. The markers show up as plus signs, and the line is dashed, as you might expect.

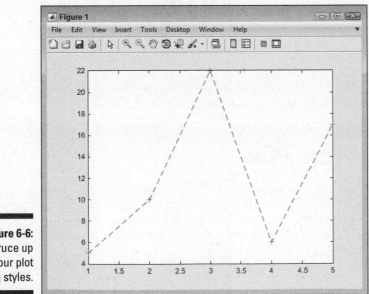

Figure 6-6: Spruce up your plot using styles.

Creating multiple plots in a single command

In many cases, you need to plot more than one set of data points when working with a plot. The `plot()` function can accommodate as many series as needed to display all your data. For example, you might want to plot both sine and cosine of x to compare them. To perform this task, you type **plot(x, sin(x), 'g-', x, cos(x), 'b-')** and press Enter (remember that x was defined earlier as `x = -pi:0.01:pi;`). Figure 6-7 shows the result.

In this case, sine appears as a green solid line. The value of cosine is in blue with a dashed line. Notice that each series appears as three values: x axis, y axis, and format string. You can add as many series as needed to complete your plot.

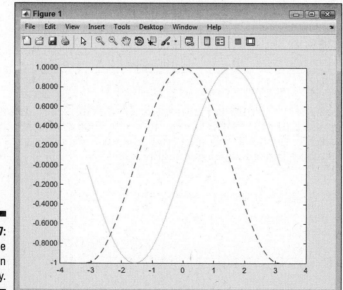

Figure 6-7:
Plot multiple series when necessary.

Modifying Any Plot

At some point, you'll want to change the content of your plot. Perhaps you want to add a legend or change how the data is presented. After you get the data looking just right, you might need to label certain items or perform other tasks to make the output look nicer. You can modify any plot you create using either commands or the MATLAB GUI.

The modification method that you use is entirely up to you. Some people work better at the keyboard, others using the mouse, and still others using a combination of the two. Working at the keyboard is a lot faster but requires that you memorize the commands to type. The GUI provides you with great memory aids, but working with the mouse is slower and you might not be able to find a particular property you want to change when it becomes buried in a menu somewhere. The following sections describe techniques to use for modifying any plot.

Making simple changes

There are a number of simple changes you can make to your plot that don't require any special handling other than to type the command. For example, to add a grid to an existing plot, you simply type **grid on** and press Enter. (MATLAB has a number of grid commands. For example, grid MINOR toggles the minor grid lines.)

Adding a legend means typing a name for each of the plots. For example, if you want to add a legend to the plot in Figure 6-7, you type **legend('Sine', 'Cosine')** and press Enter. You can also change items such as the legend orientation. The default orientation is vertical, but you can change it to horizontal by typing **legend('orientation', 'horizontal')** and pressing Enter. Notice that the property name comes first, followed by the property value.

MATLAB also lets you add titles to various parts of the plot. For example, to give the plot a title, type **title('Sine and Cosine')** and press Enter. You can also provide labels for the x-axis using xlabel() and for the y-axis using ylable(). The point is that you have full control over the appearance of the plot. Figure 6-8 shows the effects of the commands that you have tried so far. (Compare it to Figure 6-7.)

If you make a mistake, you can always clear the current plot by using the clf command. The clf command does for the plot what the clc command does for the Command window. Make sure that you actually want to clear the plot before using the clf command because there isn't any sort of undo feature to restore the plot.

Adding to a plot

You may decide that you want to add another plot to an existing plot. For example, you might want to plot the square of x for each of the values used in the previous examples. To make this technique work, you need to perform the three-step process described here:

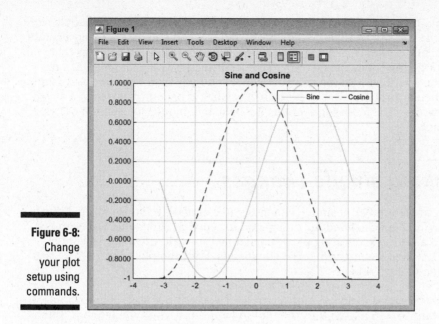

Figure 6-8:
Change
your plot
setup using
commands.

1. **Type** hold on **and press Enter.**

   If you try to add another plot without placing a hold on the current plot, MATLAB simply creates a new plot and gets rid of the old one. The hold command lets you retain the current plot while you add something to it.

2. **Type** newplot = plot(x, power(x, 2), 'm:') **and press Enter.**

   This command creates a new plot and places a handle to that plot in newplot. A *handle* is just what it sounds like — a means of obtaining access to the plot you just created. If you don't store the plot handle, you can't access it later. The output now looks like Figure 6-9.

   Notice that the legend hasn't updated itself to show the new plot. To update the legend, you must issue another legend() function call. The sine and cosine still have the same values, but the new plot has much larger values, so it appears that the previous plot lines have shrunk. However, compare the values in Figures 6-8 and 6-9 and you see that the values of sine and cosine are the same.

3. **Type** hold off **and press Enter.**

   The hold off command releases the plot. To create new plots, you must release your hold on the existing plot.

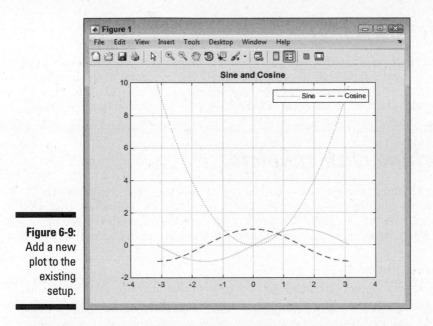

Figure 6-9:
Add a new plot to the existing setup.

Using the figure() function

This chapter concentrates on various sorts of plots because plots provide you with output. However, the figure() function can be an important part of your toolbox when you start creating scripts. You use the figure() function alone to create a new figure that doesn't have any sort of information in it. The advantage is that you can then fill the new figure with anything you want. In addition, the figure() function creates a new figure without overwriting the old one. The figure() function returns a handle to the figure rather than to the plot inside the figure. If you have multiple plots inside a figure, you can use the figure handle to select all the plots rather than just one of them.

You use the figure() function with a handle to make the figure associated with a particular handle the current figure. For example, the figure(MyFigure) command would make the figure pointed to by MyFigure the current figure. When working with multiple figures, you need some method of selecting between them, and the figure() function provides the best method of doing that.

Of course, you might have created the figure as a plot rather than as a figure. The plot handle doesn't work with the figure() function. Use the gcf() (*Get Current Figure*) function to obtain the figure handle for any figure you create using a plot. You can then save the figure handle in a variable for later use.

Deleting a plot

You might decide that you really don't want to keep a plot you've added. In this case, you need a handle to the plot you want to remove, such as the handle stored as part of the steps in the previous section. To remove the plot, type **delete(newplot)** and press Enter. MATLAB removes the plot from the display.

Working with subplots

Figure 6-9 shows three plots — one on top of the other. You don't have to display the plots in this manner. Instead, you can display them side by side (or even in a grid). To make this happen, you use the subplots feature of MATLAB. A *subplot* is simply a plot that takes up only a portion of the display.

Creating a subplot

The best way to understand subplots is to see them in action. The following steps help you create the three previous plots as subplots:

1. **Type** clf **and press Enter.**

   MATLAB clears any previous plot you created.

2. **Type** subplot(1, 3, 1) **and press Enter.**

   This function creates a grid consisting of one row and three columns. It tells MATLAB to place the first plot in the first space in the grid. You see the blank space for the plot, as shown in Figure 6-10.

3. **Type** p1 = plot(x, sin(x), 'g-') **and press Enter.**

   You see the first plot added to the display, as shown in Figure 6-11.

   Notice that the example is creating the plots one at a time. You can't combine plots in a single call when using subplots. In addition, you need to maintain a handle to each of the plots in order to configure them

4. **Type** subplot(1, 3, 2) **and press Enter.**

   MATLAB selects the second area for the next plot.

5. **Type** p2 = plot(x, cos(x), 'b-') **and press Enter.**

   You see the second plot added to the display.

6. **Type** subplot(1, 3, 3) **and press Enter.**

   MATLAB selects the third area for the next plot.

7. **Type** p3 = plot(x, power(x, 2), 'm:') **and press Enter.**

   You see the third plot added to the display, as shown in Figure 6-12.

Each plot takes up the entire area. You can't compare plots easily because each plot is in its own space and uses its own units of measure. However, this approach does have the advantage of letting you see each plot clearly.

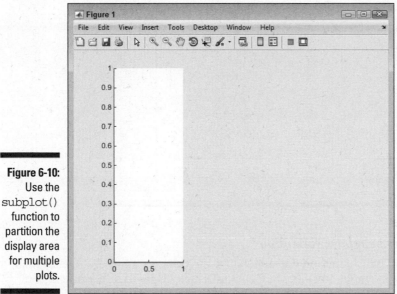

Figure 6-10:
Use the
subplot()
function to
partition the
display area
for multiple
plots.

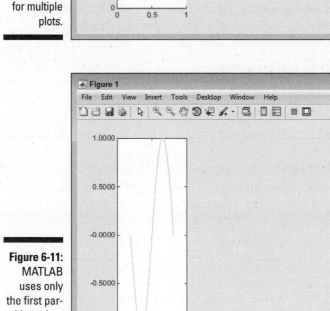

Figure 6-11:
MATLAB
uses only
the first par-
tition when
creating
the plot.

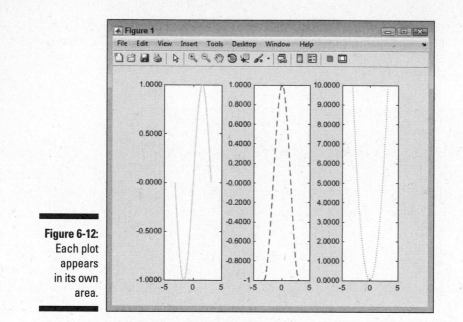

Figure 6-12:
Each plot
appears
in its own
area.

Changing subplot information

The subplot() function doesn't change anything — it merely selects something. For example, the plots in Figure 6-12 lack titles. To add a title to the first plot, follow these steps:

1. **Type** subplot(1, 3, 1) **and press Enter.**

   MATLAB selects the first subplot.

2. **Type** title('Sine') **and press Enter.**

   You see a title added to the first subplot, as shown in Figure 6-13.

Configuring individual plots

To work with a subplot in any meaningful way, you need to have a handle to the subplot. The following steps describe how to change the color and line type of the second plot:

1. **Type** subplot(1, 3, 2) **and press Enter.**

   MATLAB selects the second subplot. Even though the handle used with the set() command in the following step will select the subplot for you, this step is added so that you can actually see MATLAB select the

subplot. In some cases, performing this task as a separate step is helpful to ensure that any function calls that follow use the correct subplot,
even when these function calls don't include a handle. Later, when you
start creating scripts, you find that errors creep into scripts when you're
making assumptions about which plot is selected, rather than knowing
for sure which plot is selected.

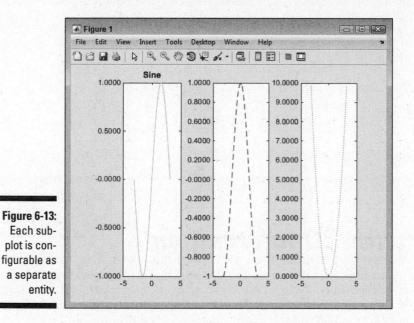

Figure 6-13:
Each sub-
plot is con-
figurable as
a separate
entity.

2. **Type** set(p2, 'color', 'r') **and press Enter.**

   The line color is now red. The `set()` function accepts a handle to a plot
   or another MATLAB object as the first value, the name of a property as
   the second, and the new value for that property as the third. This function call tells MATLAB to change the `color` property of the line pointed
   at by `p2` to red.

3. **Type** set(p2, 'LineStyle', '-.') **and press Enter.**

   The `LineStyle` property for the Cosine plot is now set to dash dot, as
   shown in Figure 6-14.

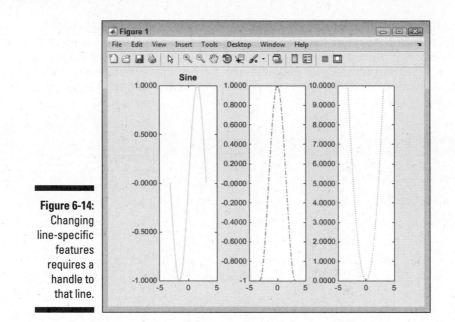

Figure 6-14:
Changing
line-specific
features
requires a
handle to
that line.

Plotting with 2D Information

MATLAB has built-in plotting routines that are suitable for many types of data and applications. Table 6-2 gives you an overview of various 2D plotting functions, including what they plot and how they're commonly used. You use these functions in place of the plot() function used throughout the chapter to create plots. The output will contain the kind of plot you have requested, such as a pie chart when using the pie() function. (MATLAB also supports 3D plotting; for more on that aspect of plotting, check out Chapter 7.)

Table 6-2	MATLAB Plotting Routines	
Routine	*What It Plots*	*Used By*
plotyy()	Data with two y axes	Business users rely on this plot to show two sets of units, for example, quantity sold and money.
loglog()	Data with both x and y axes as log scales	Science, Technology, Engineering, and Mathematics (STEM) users rely on this plot to show power or root dependence of y versus x.

Routine	What It Plots	Used By
semilogx()	Data with x axis log scale	STEM users rely on this plot to show logarithmic dependence of y versus x.
semilogy()	Data with y axis log scale	STEM and social science users rely on this plot to show exponential dependence of y versus x and population growth (as an example).
scatter()	Data in x-y pairs	Experimentalists and statisticians rely on this plot to show patterns created by the individual data points.
hist()	Frequency of occurrence of particular values of data	Experimentalists and statisticians rely on this plot to understand imprecision and inaccuracy.
area()	x-y data with areas filled in	Business and STEM users rely on this plot to see (and understand) the contributions of parts to a whole.
pie()	Set of labeled numbers	Business users rely on this plot to see (and understand) the fractional contributions of each part to a whole.
ezpolar()	Data in terms of radius and angle	STEM users rely on this plot to show the angular dependence of information.

Chapter 7

Using Advanced Plotting Features

In This Chapter

▶ Working with 3D plots

▶ Creating enhanced plots

Chapter 6 helps you create plots that convey 2D data in visual form. Using plots in this manner helps you present the data in a way that most humans understand better than abstract numbers. Visual presentations are concrete and help the viewer see patterns that might be invisible otherwise. The 3D plots described in this chapter do the same thing as those 2D plots, only with a 3D data set. The viewer sees depth as well as height and width when looking at that data. Using a 3D data set can greatly improve the amount of information the user obtains from a plot. For example, a 3D plot could present the variation of a data set over time so that the user gains insights into how the data set changes.

If you worked through Chapter 6, you focused mostly on small changes to improve the aesthetics of your plots. This chapter looks at some of the fancier things you can do to make plots even more appealing. In many cases, nontechnical viewers require these sorts of additions in order to appreciate the data you present. Making data as interesting as possible can only help to improve your presentation and convince others to accept your interpretation of the data. Of course, making plots that look nice is also just plain fun, and everyone could use a little more fun in their creation and presentation of data.

This chapter focuses on the kinds of 3D plots that you perform most often. Appendix B provides a more comprehensive listing of the plot types that MATLAB supports. Be sure to also check out the online materials for this book (as described in the Introduction) and the blog posts at http://blog. johnmuellerbooks.com to see how to work with other plot types. MATLAB provides so much in the way of plot functionality that you could possibly create a book on just that one topic.

Plotting with 3D Information

A 3D plot has an x, y, and z axis (height, width, and depth, if you prefer). The addition of depth lets you present more information to the viewer. For example, you could present historical information about a plot so that each element along the z axis is a different date. Of course, the z axis, like the x and y axes, can represent anything you want. The thing to remember is that you now have another method of presenting information to the viewer.

It's also important to consider that you're presenting 3D information on a 2D surface — the computer screen or a piece of paper. Some users forget this fact and find that some of their data hides behind another plot object that is greater in magnitude. When working with 3D plots, you need to arrange the information in such a manner that you can see it all onscreen.

The following sections describe various kinds of plots and how to create them. Each plot type has specific uses and lends itself to particular kinds of data display. Of course, the kind of plot you choose depends on how you want to present the data as well.

Using the bar () function to obtain a flat 3D plot

The bar chart is a standard form of presentation that is mostly used in a business environment. You can use a bar chart to display either 2D or 3D data. When you feed a bar chart a vector, it produces a 2D bar chart. Providing a bar chart with a matrix produces a 3D chart. The following steps help you create a 3D bar chart.

1. **Type** SurveyData = [8, 7, 6; 13, 21, 15; 32, 27, 32] **and press Enter.**

    MATLAB creates a new matrix named `SurveyData` that is used for many of the examples in this chapter. You see the following output:

    ```
    SurveyData =
         8      7      6
        13     21     15
        32     27     32
    ```

2. **Type** bar(SurveyData) **and press Enter.**

    You see a flat presentation of `SurveyData`, as shown in Figure 7-1. The x axis shows each of the columns in turn. (If you could see colors in the book, you would see that the first column is blue, the second is green, and the third is red.) The y axis presents the value of each cell (such as 8, 7, and 6 for the first `SurveyData` row). The z axis presents each row in a group, and each group corresponds to a number between 1 and 3.

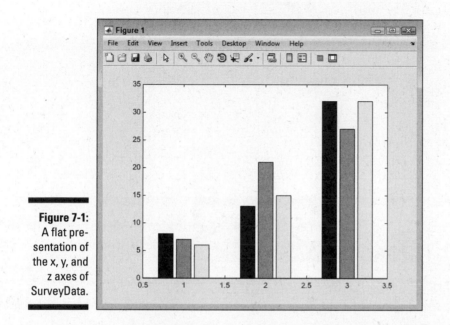

Figure 7-1:
A flat presentation of the x, y, and z axes of SurveyData.

3. **Type** Bar1 = bar(SurveyData, 'stacked') **and press Enter.**

You see the same `SurveyData` matrix presented as a stacked bar chart, as shown in Figure 7-2. In this case, the x axis elements are shown stacked one on top of the other.

The example also outputs information about the bar chart handles (a means of obtaining access to the plot). The values may differ, but you should see three handles output like the following (each handle is named Bar — previous versions of MATLAB used a number to represent the handle in the output):

```
Bar1 =
  1x3 Bar array:

    Bar    Bar    Bar
```

Each of the z axis elements has its own handle that you use to manipulate it. This is an important part of working with the bar chart later when you want to modify something.

Figures 7-1 and 7-2 present two forms of the same data. The `bar()` function provides you with several alternative presentations:

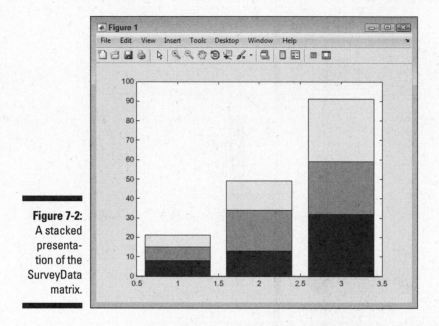

Figure 7-2:
A stacked presentation of the SurveyData matrix.

- `grouped`: This is the default setting shown in Figure 7-1.

- `hist`: The data appears much like in Figure 7-1, except that no spaces appear between the bars for a particular group. The groups do still have spaces between them.

- `hisc`: The groups are positioned so that each group starts at a number on the x axis, rather than being centered on it.

- `stacked`: This is the stacked appearance shown in Figure 7-2.

4. Type get(Bar1(1)) **and press Enter.**

The `get()` function obtains the properties you can work with for a particular object. In this case, you request Bar1(1), which is the first group in Figure 7-2. In other words, this would be the first member of the z axis. You see the following output:

```
        Annotation: [1x1 matlab.graphics.
eventdata.Annotation]
         BarLayout: 'stacked'
          BarWidth: 0.8000
          BaseLine: [1x1 Baseline]
```

```
             BaseValue: 0
         BeingDeleted: 'off'
           BusyAction: 'queue'
       ButtonDownFcn: ''
             Children: []
             Clipping: 'on'
            CreateFcn: ''
            DeleteFcn: ''
          DisplayName: ''
            EdgeColor: [0 0 0]
            FaceColor: 'flat'
     HandleVisibility: 'on'
              HitTest: 'on'
           Horizontal: 'off'
        Interruptible: 'on'
            LineStyle: '-'
            LineWidth: 0.5000
               Parent: [1x1 Axes]
             Selected: 'off'
   SelectionHighlight: 'on'
         ShowBaseLine: 'on'
                  Tag: ''
                 Type: 'bar'
        UIContextMenu: []
             UserData: []
              Visible: 'on'
                XData: [1 2 3]
            XDataMode: 'auto'
          XDataSource: ''
                YData: [8 13 32]
          YDataSource: ''
```

After you know the properties that you can modify for any MATLAB object, you can use those properties to start building scripts. (You created your first script in Chapter 2.) Just creating and then playing with objects is a good way to discover just what MATLAB has to offer. Many of these properties will appear foreign to you and you don't have to worry about them, but notice that the YData property contains a vector with the three data points for this particular bar.

It's also possible to obtain individual property values. For example, if you use the get(Bar1(1), 'YData') command, you see the current YData values for just the first bar.

5. **Type** set(Bar1(1), 'YData', [40, 40, 40]) **and press Enter.**

The set() function lets you modify the property values that you see when using the get() function. In this case, you modify the YData property for the first bar — the blue objects when you see the plot on screen. Figure 7-3 shows the result of the modification.

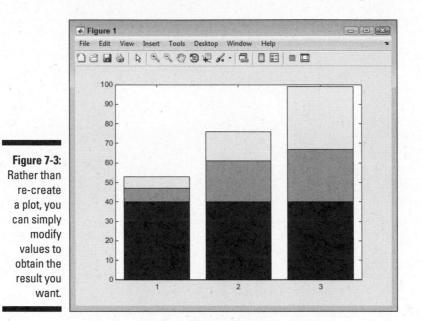

Figure 7-3:
Rather than
re-create
a plot, you
can simply
modify
values to
obtain the
result you
want.

Using bar3() to obtain a dimensional 3D plot

The flat form of the 3D plot is nice, but it lacks pizzazz. When you present your information to other engineers and scientists, the accuracy of the flat version is welcome. Everyone can see the 3D data clearly and work with it productively. A business viewer might want something a bit different. In this case, presenting a pseudo-3D look is better because the business user gets a better overall view of the data. Precise measurements aren't quite as useful in this case — but seeing how the data relate to each other is. To create a dimensional plot of the data that appears in the previous section, type **Bar2 = bar3(SurveyData)** and press Enter. You see a result similar to the one shown in Figure 7-4.

The two problems with the presentation in Figure 7-4 are that you can't see some of the data and none of it is presented to best effect. In order to rotate the image, you use the view() function. The view() function can accept either x, y, and z rotation in degrees, or a combination of azimuth and elevation. Using x, y, and z rotation is easier for most people than trying to figure out azimuth and elevation. To change Figure 7-4 so that you can more easily see the bars, type **view([-45, 45, 30])** and press Enter. Figure 7-5 shows the result.

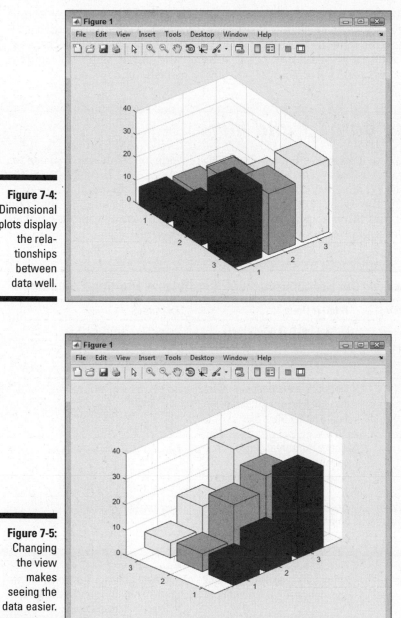

Figure 7-4:
Dimensional plots display the relationships between data well.

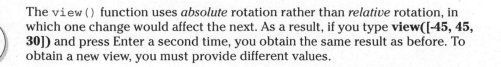

Figure 7-5:
Changing the view makes seeing the data easier.

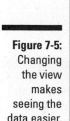

The view() function uses *absolute* rotation rather than *relative* rotation, in which one change would affect the next. As a result, if you type **view([-45, 45, 30])** and press Enter a second time, you obtain the same result as before. To obtain a new view, you must provide different values.

As an alternative to using the view() function, you can also click the Rotate 3D button, shown in Figure 7-5. It's the button with the circular arrow that appears to the right of the hand icon. Although the view() function is more precise and lets you make changes to the view without moving your hands from the keyboard, the Rotate 3D button can be faster and easier.

Using barh() and more

MATLAB provides you with a number of 3D plotting functions that you use to obtain various effects. The barh(), bar3(), and bar3h() functions work just like the bar() function except that they display slightly differently. Closely related are the hist(), histc(), rose(), polar(), and pareto() functions. Table 7-1 lists the various plotting functions that you have at your disposal and a brief description of how they work.

Table 7-1   Bar Procedures and Other Related Plotting Procedures

Function	What It Does	Examples
bar()	Plots a flat bar chart that relies on color and grouping to show the z axis.	bar(SurveyData) bar(SurveyData', 'stacked')
bar3()	Plots a dimensional bar chart that uses color and perspective to show the z axis.	bar3(SurveyData) bar3(SurveyData', 'stacked')
bar3h()	Plots a horizontal dimensional bar chart that uses color and perspective to show the z axis.	bar3h(SurveyData) bar3h(SurveyData', 'stacked')
barh()	Plots a horizontal flat bar chart that relies on color and grouping to show the z axis.	barh(SurveyData) barh(SurveyData', 'stacked')
hist()	Plots frequency of occurrence for bins given raw data and, optionally, bin centers.	hist(randn(1,100), 5) % creates 100 normally distributed random numbers and places them in five equally spaced bins. hist(randn(1,100), [-3.5,-2.5,-1.5,-.5, .5,1.5,2.5,3.5]) % creates 100 normally distributed numbers and places them in specific bin centers.

Function	What It Does	Examples
`histc()`	Obtains frequency data for each bin and displays it as text (rather than as a plot). The advantage is that you can specify bin edges.	`histc(randn(1,100), [-4:1:4])` % specifies bins that are 1 unit wide, with edges on integers starting at -4. You could use this information in a plot as `bar([-4:1:4],ans, 'histc').`

`pareto()`	Plots a bar chart ordered by highest bars first — used in business to identify factors causing the greatest effect.	`histc(ra ndn(1,100),[-4:1:4])`
		
		`pareto(ans)`

`polar()`	Plots a polar display of data in which the rings of the circle represent individual data values.	`histc(randn(1,100), [-4:1:4])`
		
		`polar(ans)`

`rose()`	Plots data bars versus angles in a polar-like display. As with the `hist()` function, you may also specify bin centers.	`rose(randn(1,100), 5)` % creates 100 normally distributed numbers and places them in five equally spaced bins.

Enhancing Your Plots

For visual information to be meaningful and more informative, you need to add titles, labels, legends, and other enhancements to plots of any type (both 3D and 2D). (The greater visual appeal of 3D plots only makes the plot prettier, not more informative.) The following sections of the chapter won't make you into a graphic designer, but they will let you create more interesting plots that you can use to help others understand your data. The goal of these sections is to help you promote better communication. The examples in the following sections rely on the 3D plot you created in the "Using bar3() to obtain a dimensional 3D plot" section, earlier in this chapter.

Getting an axes handle

Before you can do anything, you need a handle to the current axes. The best way to obtain such a handle is to type **Bar2Axes = gca()** (*G*et *C*urrent *A*xes) and press Enter. The `gca()` function returns the handle for the current plot. When you type **get(Bar2Axes)** and press Enter, you see the properties associated with the current plot.

Tricks of the trade for working with figures

Knowing a few tricks is helpful when working with plots. The tricks help you perform work faster and more efficiently. In addition, they make working with plots more fun.

✔ Start over by using the `clf` command, which stands for Clear Figure (which is precisely what it does).

✔ Stay organized by obtaining a handle (a method to gain access to the figure) using the `gcf()` (*Get Current Figure*) function. Don't confuse the figure handle with the plot handle mentioned earlier — a figure contains a plot, so the figure handle is different.

✔ Make a particular figure the current figure, use the `figure()` function with the variable containing the figure handle.

✔ Reset figures to default values using the `reset()` function with the variable containing the figure handle. This feature comes in handy when the changes you make produce undesirable results.

✔ See the properties associated with the current figure by using the `get()` function with the variable containing the figure handle. MATLAB displays a list of the properties and their current values. If a property doesn't have a value, the value is left blank.

Modifying axes labels

MATLAB automatically creates labels for some of the axes for you. However, the labels are generic and don't really say anything. To modify anything on the axes, you need an axes handle (as described in the previous section).

After you have the handle, you use the appropriate properties to modify the appearance of the axes. For example, to modify the x axis label, you type **xlabel(Bar2Axes, 'X Axis')** and press Enter. Similarly, for the y axis, you type **ylabel(Bar2Axes, 'Y Axis')** and press Enter. You can also use the `zlabel()` function for the z axis.

Each of the ticks on an axis can have a different label as well. The default is to simply assign them numbers. However, if you want to assign meaningful names to the x axis ticks, you can type **set(Bar2Axes, 'XTickLabel', {'Yesterday', 'Today', 'Tomorrow'})** and press Enter. Notice that the labels appear within a cell array using curly brackets ({}). Likewise, to set the y axis ticks, you can type **set(Bar2Axes, 'YTickLabel', {'Area 1', 'Area 2', 'Area3'})** and press Enter. You can also use a `ZTickLabel` property, which you can modify.

To control the tick values, you type **set(Bar2Axes, 'ZTick', [0, 5, 10, 15, 20, 25, 30, 35, 40])** and press Enter. Those two axes also have `XTick` and `YTick` properties. Of course, in order to see the z axis ticks, you also need to change the limit (the size of the plot in that direction). To perform this task you type **set(Bar2Axes, 'ZLim', [0 45])** and press Enter.

Many of the `set()` function commands have alternatives. For example, you can change the `ZLim` property by using the `zlim()` function. The alternative command in this case is `zlim(Bar2Axes, [0 45])`. Using a `set()` function does have the advantage of making it easier to enter the changes because you have to remember only one function name. However, the result is the same no matter which approach you use, so it's entirely a matter of personal preference.

Use the `get()` function whenever necessary to discover additional interesting properties to work with. Properties are available to control every aspect of the axes' display. For example, if you want to change the color of the axes' labels, you use the `XColor`, `YColor`, and `ZColor` properties. Figure 7-6 shows the results of the changes in this section.

Many properties have an automatic setting. For example, to modify the `ZLim` property so that it uses the automatic setting, you type **zlim(Bar2Axes, 'auto')** and press Enter. The alternative when using a `set()` function is to type **set(Bar2Axes, 'ZLimMode', 'auto')** and press Enter. Notice that when you use the `zlim()` function, you can set either the values or the mode using the same command. When using the `set()` function, you use different properties (`ZLim` and `ZLimMode`) to perform the task. However, the important thing to remember is that the `auto` mode tells MATLAB to configure these items automatically for you.

Using commands to change plot properties is fast and precise because your hands never leave the keyboard and you don't spend a lot of time searching for a property to change in the GUI. However, you can always change properties using the GUI as well. Click the Edit Plot button (the one that looks like a hollow arrow on the left side of the magnifying glasses in Figure 7-5) to put the figure into edit mode. Click the element you wish to modify to select it. Right-click the selected element and choose Show Property Editor to modify the properties associated with that particular element.

Adding a title

Every plot should have a title to describe what the plot is about. You use the `title()` function to add a title. However, the `title()` function accepts all sorts of properties so that you can make the title look precisely the way you want. To see how this function works, type **title(Bar2Axes, 'Sample Plot', 'FontName', 'Times', 'FontSize', 22, 'Color', [.5, 0, .5], 'BackgroundColor', [1, 1, 1], 'EdgeColor', [0, 0, 0], 'LineWidth', 2, 'Margin', 4)** and press Enter. MATLAB changes the title, as shown in Figure 7-7.

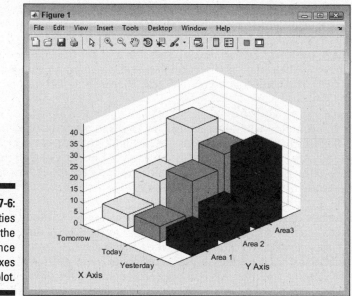

Figure 7-6:
Properties
control the
appearance
of the axes
in your plot.

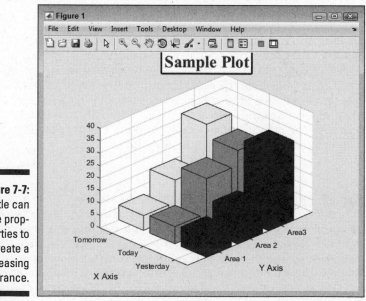

Figure 7-7:
A title can
use prop-
erties to
create a
pleasing
appearance.

Interestingly enough, most of the plot objects support these properties, but
the title uses them most often. Here's a list of the properties you just changed:

✔ `FontName`: Provides the text name of a font you want to use. It can be the name of any font that is stored on the host system.

✔ `FontSize`: Specifies the actual size of the font (in points by default). A larger number creates a larger font.

✔ `Color`: Determines the color of the text in the title. This property requires three input values for red, green, and blue. The values must be between 0 and 1. You can use fractional values and mix colors as needed to produce specific results. An entry of all zeros produces black — all ones produces white.

✔ `BackgroundColor`: Determines the color of the background behind the text in the title. It uses the same color scheme as the `Color` property.

✔ `EdgeColor`: Determines the color of any line surrounding the title. It uses the same color scheme as the `Color` property.

✔ `LineWidth`: Creates a line around the title of a particular width (in points by default).

✔ `Margin`: Adds space between the line surrounding the title (the edge) and the text (in points by default).

Rotating label text

In some cases, the text added to a plot just doesn't look right because it doesn't quite reflect the orientation of the plot itself. The title in Figure 7-7 looks just fine, but the x axis and y axis labels look slightly askew. You can modify them so that they look better.

When you review some properties using the `get()` function, you see a handle value instead of an actual value. For example, when you look at the `XLabel` value, you see a handle that lets you work more intimately with the underlying label. To see this value, you use the `get(Bar2Axes, 'XLabel')` command. If you don't want to use a variable to hold the handle, you can see the `XLabel` properties by typing **get(get(Bar2Axes, 'XLabel'))** and pressing Enter. What you're telling MATLAB to do is to get the properties that are pointed to by the `XLabel` value obtained with the `Bar2Axes` handle — essentially, a handle within a handle.

One of the properties within `XLabel` is `Rotation`, which controls the angle at which the text is displayed. To change how the plot looks, type **set(get(Bar2Axes, 'XLabel'), 'Rotation', -30)** and press Enter. The x axis label is now aligned with the plot. You can do the same thing with the y axis label by typing **set(get(Bar2Axes, 'YLabel'), 'Rotation', 30)** and pressing Enter.

You can also reposition the labels, although using the GUI to perform this task is probably easier. However, the `Position` property provides you with access to this feature. To see the starting position of the x axis label, type **get(get(Bar2Axes, 'XLabel'), 'Position')** and press Enter. The example setup shows the following output:

```
ans =
    1.4680    -1.3631            0
```

Small tweaks work best. Type **set(get(Bar2Axes, 'XLabel'), 'Position', [1.50 -1.3 1])** and press Enter to better position the x axis label. (You may need to fiddle with the numbers a bit to get your plot to match the one in the book, and your final result may not look precisely like the screenshot.) After a little fiddling, your X Axis label should look like the one in Figure 7-8.

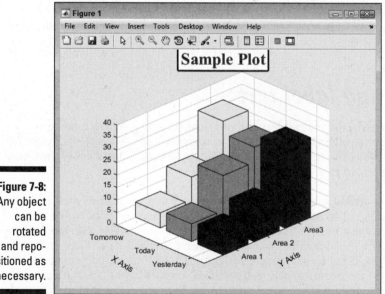

Figure 7-8: Any object can be rotated and repositioned as necessary.

Employing annotations

Annotations let you add additional information to a plot. For example, you might want to put a circle around a particular data point or use an arrow to point to a particular bar as part of your presentation. Of course, you may simply want to add some sort of emphasis to the plot using artistic elements. No matter how you want to work with annotations, you have access to these drawing elements:

✔ Line

✔ Arrow

✔ Text Arrow

✔ Double Arrow

✔ Textbox

✔ Rectangle

✔ Ellipse

To add annotations to your figure, you use the `annotation()` function. Say you want to point out that Area 3 in Figure 7-8 is the best area of the group. To add the text area, you type **TArrow = annotation('textarrow', [.7, .55], [.9, .77], 'String', 'Area 3 is the best!')** and press Enter. You see the result shown in Figure 7-9. This version of the `annotation()` function accepts the annotation type, the x location, y location, property name (`String`), and property value (`Area 3 is the best!`).

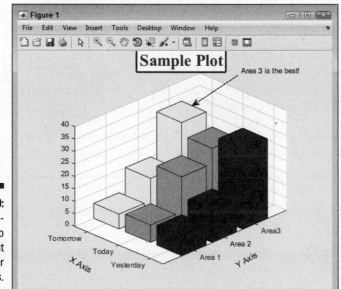

Figure 7-9:
Add annotations to document your plot for others.

The annotations don't all use precisely the same command format. For example, when you want to add a textbox, you provide the starting location, height, and width, all within the same vector. To see this version of the `annotation()` function in action, type **TBox = annotation('textbox', [.1, .8, .11, .16], 'String', 'Areas Report', 'HorizontalAlignment', 'center', 'VerticalAlignment', 'middle')** and press Enter. In this case, you center the text within the box and place it in the upper-left corner. A textbox doesn't point to anything — it simply displays information, as shown in Figure 7-10.

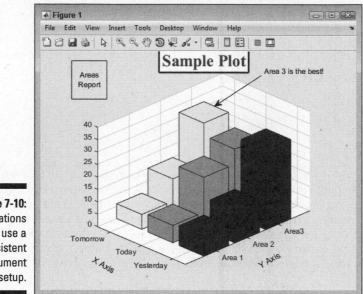

Figure 7-10:
Annotations
don't use a
consistent
argument
setup.

Printing your plot

At some point, you probably need to print your plot. You have a number of choices in creating output. The following list provides you with a quick overview of the options at your disposal:

✔ At the Figure window, select File➪Print to display the Print dialog box — select the options you want to use for printing.

✔ At the Figure window, type Ctrl+P to display the Print dialog box — select the options you want to use for printing.

✔ At the Command window, type **print()** and press Enter.

• Using `print()` alone prints the entire figure, including any subplots.

• Adding a handle to `print()`, such as `print(Bar2)`, prints only the object associated with the handle.

In some cases, you may want to output your plot in a form that lets you print it in another location. When working in the Figure window, you select the Print to File option in the Print dialog box. MATLAB will ask you to provide a filename for printing. When working in the Command window, you supply a filename as a second argument to the `print()` function. For example, you might use `print(Bar2, 'MyFile.prn')` as the command.

Part III
Streamlining MATLAB

"SayHello2" requires more input arguments to run:

SayHello2(Name)

Enter input arguments by typing them now, then press Enter to run. The values you enter will be set as the default inputs when you click the Run button in the future.

Run and Advance

Run: SayHello2(Name)

Run: *type code to run*

See an example of how to use an anonymous function to provide input for a contour plot at http://www.dummies.com/extras/matlab.

In this part . . .

- ✔ Create scripts to automate tasks.
- ✔ Work with functions when performing complex tasks.
- ✔ Discover the uses of inline and anonymous functions.
- ✔ Use comments to document scripts and functions.
- ✔ See how scripts and functions can make decisions.
- ✔ Develop scripts and functions that perform tasks more than one time.

Chapter 8

Automating Your Work

*G*etting the computer to do the work for you is probably one of the best reasons to use a computer in the first place. Anytime you can automate repetitive or mundane tasks, you free yourself to do something more interesting. MATLAB is an amazing tool for performing all sorts of creative work, but you also have a lot of mundane and repetitive tasks to perform. For example, you may need to generate the same plot every week for a report. Automating that task would free you to do something more interesting, such as discover a cure for cancer or send a rocket to Mars. The point is that you have better things to do with your time, and MATLAB is only too willing to free your time so that you can do them. That's what scripting is all about — it isn't about being some mad genius geek, it's all about automating tasks so that you can do something more interesting.

Scripting is simply a matter of writing a *procedure* — writing down exactly what you want the computer to do for you, in other words. (It's also possible to compare it to a making a movie, where a writer creates the words the actors say and specifies the actions they perform.) You likely write procedures all the time for various people in your life. In fact, you may write procedures for yourself so that you remember how to perform the task later. Scripting for MATLAB is no different from any other procedure you have written in the past, except that you need to write the procedure in a manner that MATLAB understands.

This chapter helps you create basic MATLAB scripts, save them to disk so that you can access them whenever you want, and then run the scripts as needed. You also discover how to make your scripts run fast so that you don't have to wait too long for MATLAB to complete its work. Finally, this chapter helps you understand the nature of errors in scripts, and how to locate and fix them.

Understanding What Scripts Do

A script is nothing more than a means to write a procedure that MATLAB can follow to perform useful work. It's called a script and not a procedure because a script follows a specific format. MATLAB actually speaks its own English-like language that you must use to tell it what to do. The interesting thing is that you've used that language in every chapter so far. A script doesn't do much more than link together the various commands that you have used to perform a task from one end to the other. The following sections describe what a script does in more detail.

Creating less work for yourself

The object of a script is to reduce your workload. This concept might seem straightforward now, but some people get so wrapped up in the process of creating scripts that they forget that the purpose of the script is to create less work, not more. In fact, a script should meet some (or with luck, all) of the following goals:

- ✔ Reduce the time required to perform tasks
- ✔ Reduce the effort required to perform tasks
- ✔ Allow you to pass the task along to less skilled helpers
- ✔ Make it possible to perform the tasks with fewer errors (the computer will never get bored or distracted)
- ✔ Create standardized and consistent output
- ✔ Develop a secure environment in which to perform the task (because the details are hidden from view)

Notice that none of the goals in this list entail making the computer do weird things that it doesn't normally do or wasting your time writing scripts to perform tasks that you never did in the past. The best scripts perform tasks that you already know how to do well because you have performed them so many times in the past. Yes, it's entirely possible that you could eventually create a script to perform a new task, but even in that case, the new task is likely built on tasks that you have performed many times in the past. Most people get into trouble with scripting when they try to use it for something they don't understand or haven't clearly defined.

Defining when to use a script

Scripts work well only for mundane and repetitive tasks. Sometimes writing a script is the worst possible thing you can do. In fact, many times you can find yourself in a situation in which writing a script causes real (and potentially irreparable) damage. The following list provides you with guidelines as to when to use a script:

- ✔ The task is repeated often enough that you actually save time by writing a script (the time saved more than offsets the time spent writing the script).

- ✔ The task is well defined, so you know precisely how to perform it correctly.

- ✔ There are few variables in the way in which the task is performed so that the computer doesn't have to make many decisions (and the decisions it makes are from a relatively small set of potential absolute answers).

- ✔ No creativity or unique problem-solving abilities are required to perform the task.

- ✔ All the resources required to perform the task are accessible by the host computer system.

- ✔ The computer can generally perform the task without constantly needing to obtain permissions.

- ✔ Any input required by the script is well defined so that the script and MATLAB can understand (and anticipate) the response.

Believe it or not, you likely perform regularly a huge number of tasks that fulfill all these requirements. The important thing is to weed out those tasks that you really must perform by yourself. Automation works only when used correctly to solve specific problems.

Creating a Script

Creating a script can involve nothing more than writing commands. In fact, the sections that follow show a number of ways in which you can create simple scripts without knowing anything about scripting. It may even strike you as quite odd that scripting feels much like writing commands in the Command window. The only difference is that the commands don't execute immediately. That's the point of these following sections: Scripting doesn't have to be hard or complicated; it only needs to solve the problems you normally solve anyway.

Writing your first script

MATLAB provides many different ways to write scripts. Some of them don't actually require that you write anything at all! However, the traditional way to create a script in any application is to write it, so that's what this first section does — shows you how to write a tiny script. The most common first script in the entire world is the "Hello World" example. The following steps demonstrate how to create such a script using MATLAB.

1. **Click New Script on the Home tab of the menu.**

   You see the Editor window appear, as shown in Figure 8-1. This window provides the means to interact with scripts in various ways. The Editor tab shown in the figure is the one you use most often when creating new scripts.

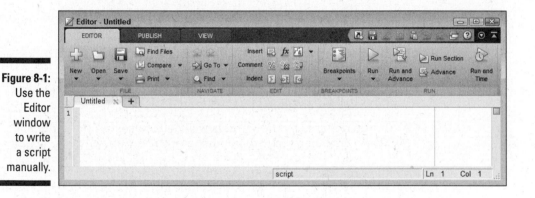

Figure 8-1:
Use the Editor window to write a script manually.

2. **Type** 'Hello World'.

   The text is highlighted in a light orange, and a squiggly red line appears under it. When you hover your mouse over the squiggly line, you see the message shown in Figure 8-2.

   In this case, you ignore the error because you want to see the output. However, if you wanted to correct the problem (the way MATLAB thinks you should), you could either type a semicolon or click Fix to resolve the issue. MATLAB will always tell you if it thinks that you're making a mistake, but sometimes MATLAB is overzealous (as in this situation).

3. **Click Run on the Editor tab of the Editor window.**

   You see a Select File for Save As dialog box, as shown in Figure 8-3. MATLAB always requests that you save your script before you run it to ensure that your script doesn't get lost or corrupted in some way should something happen when it runs.

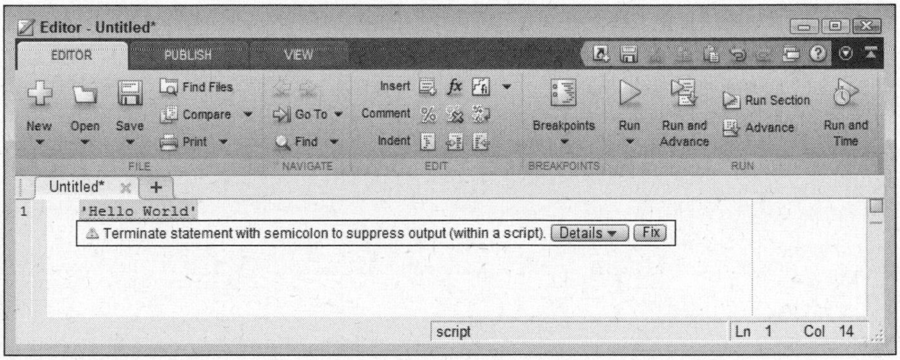

Figure 8-2:
The Editor
tells you
when it
thinks that
you're
making a
mistake.

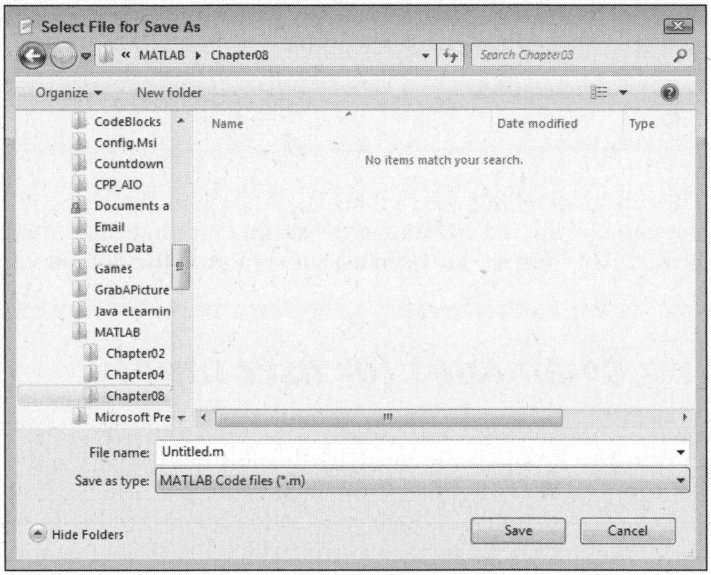

Figure 8-3:
MATLAB
always asks
you to save
your work
before you
run a script.

4. **Create or select the** `MATLAB\Chapter08` **directory, type** `FirstScript.m`
   in the File Name field, and click Save.

   MATLAB saves your script to disk. All your script files will have an `.m`
   extension.

   At this point, you may see a MATLAB Editor dialog box appear like the
   one shown in Figure 8-4. If that happens, simply click the box's Add to
   Path button to make the dialog box disappear. If you don't see this box,
   continue to Step 5.

Figure 8-4:
The direc-
tory you use
to store the
script must
be the cur-
rent direc-
tory or in the
MATLAB
path.

MATLAB Editor

(i) **MATLAB cannot run this file because
C:\MATLAB\Chapter02\FirstScript.m shadows it in the MATLAB path.**

To run this file, you can either change the MATLAB current folder or add its
folder to the MATLAB path.

[Change Folder] [Add to Path] [Cancel] [Help]

5. Select the MATLAB window.

You see the following script output:

```
>> FirstScript
ans =
Hello World
```

The output is telling you that MATLAB has run `FirstScript`, which is
the name of the file containing the script, and that the output is `Hello
World`. This output has been assigned to `ans`, the default variable.

Using commands for user input

Some scripts work just fine without any user input, but most don't. In order to
perform most tasks, the script must ask the user questions and then react to
the user's input. Otherwise, the script must either perform the task precisely
the same way every time or obtain information from some other source. User
input makes it possible to vary the way in which the script works.

Listing 8-1 shows an example of a script that asks for user input. You can
also find this script in the `AskUser.m` file supplied with the downloadable
source code.

Listing 8-1:   Asking for User Input

```
Name = input('What is your name? ', 's');
disp(['Hello ', Name]);
```

The `input()` function asks for user input. You provide a prompt that tells
the user what to provide. When you want string input, as is the case in this
example, you add the `'s'` argument to tell MATLAB that you want a string
and not a number. When the user types a name and presses Enter, the value
is placed in `Name`.

The disp() function outputs text without assigning it to a variable. However, the disp() function accepts only a single input and the example needs to output two separate strings (the "Hello" part and the "*Name*" part) as a combined whole. To fix this problem, you use the concatenation operator ([]). The term *concatenation* simply means to combine two strings. You separate each of the strings with a comma, as shown in the example.

When you run this example, the script asks you to type your name. Type *your name* and press Enter. In this case, the example uses John as the name, but you can use any name you choose. After you press Enter, the script outputs the result. Here is typical output from this example:

```
>> AskUser
What is your name? John
Hello John
```

Copying and pasting into a script

Experimentation is an essential part of working with MATLAB. After you get a particular command just right, you may want to add it to a script. This act involves cutting and pasting the information. When working in the Command window, simply highlight the text you want to move into a script, right-click it, and choose Copy or Cut from the context menu. As an alternative, most platforms support speed keys for cutting and pasting, such as Ctrl+C for copy and Ctrl+X for Cut.

Copying and cutting places a copy of the material on the Clipboard. Select the Editor window, right-click the location where you want to insert the material, and choose Paste from the context menu. (The pasted material is always put wherever the mouse pointer is pointing, so make sure you have the mouse cursor in the right place before you right click.) As an alternative, most platforms provide a speed key for pasting, such as Ctrl+V. In this case, you place the insertion pointer (the text pointer) where you want the new material to appear.

The Command History window succinctly stores all the commands that you type, making it easy for you to pick and choose the commands you want to place in a script. The following list provides techniques that you can use in the Command History window:

- ✔ Click a single line to use just that command.
- ✔ Ctrl+Click to add additional lines to a single line selection.
- ✔ Shift+Click to add all the lines between the current line and the line you clicked to a single line selection.

The result is that you end up with one or more selected lines. You can cut or copy these lines to the Clipboard and then paste them into the Editor window.

Using other sources for script material is possible, and you should use them whenever you can. For example, when you ask for help from MATLAB, the help information sometimes includes example code that you can copy and paste into your script. You can also find online sources of scripts that you can copy and paste. Checking the results of the pasting process is important in this case to ensure that you didn't inadvertently copy nonscript material. Simply delete the unwanted material before you save the script.

Converting the Command History into a script

After experimenting for a while, you might come up with a series of commands that does precisely what you'd like that series to do. Trying to cut and paste the commands from the Command window is inconvenient. Of course, you could select the commands in the Command History window, copy them to the clipboard, and paste them from there, but that seems like a waste of time too.

In reality, you can simply make a script out of the commands that you select in the Command History window. After you select the commands you want to use, just right-click the selected commands and choose Create Script from the context menu that appears. MATLAB opens a new Editor window with the selected commands in place (in the order they appear in the Command History window). Save the result to disk and run the script to see how it works.

Continuing long strings

Sometimes you can't get by with a short prompt — you need a longer prompt in order to obtain the information you need. When you need to create a longer string, use the continuation operator (. . .), which many people will recognize as an ellipsis. Listing 8-2 shows an example of how you can use long strings in a prompt to modify the UserInput example shown in Listing 8-1. You can also find this script in the LongString.m file supplied with the downloadable source code.

Listing 8-2: Asking for User Input in a Specific Way

```
Prompt = ['Type your own name, but only if it isn''t ',...
          'Wednesday.\nType the name of the neighbor ',...
          'on your right on Wednesday.\nHowever, on ',...
          'a Wednesday with a full moon, type the ',...
          'name of\nthe neighbor on your left! '];
Name = input(Prompt, 's');
disp(['Hello ', Name]);
```

This example introduces several new features. The `Prompt` variable contains a long string with some formatting that you haven't seen before. It uses the concatenation operator to create a single string from each of the lines in the text. Each substring is self-contained and separated from the other substrings with a comma. The continuation operator lets you place the substrings on separate lines.

Notice the use of the double single quote (isn"t) in the text. You need to use two single quotes when you want a single quote to appear in the output as an apostrophe (isn't), rather than terminate a string. The `\n` character is new, too. This is a special character that controls how the output appears, so it is called a *control character.* In this case, the `\n` character adds a new line. When you run this example, you see output similar to that shown here:

```
LongString
Type your own name, but only if it isn't Wednesday.
Type the name of the neighbor on your right on Wednesday.
However, on a Wednesday with a full moon, type the name of
the neighbor on your left! John
Hello John
```

Everywhere a `\n` character appears in the original string, you see a new line. In addition, the word `isn't` contains a single quote, as expected. Table 8-1 shows the control characters that MATLAB supports, and defines how they are used.

Table 8-1   MATLAB Control Characters

Character Use	Character Sequence
Single quotation mark/apostrophe	''
Percent character	%%
Backslash	\\
Alarm (sounds a beep or tone on the computer)	\a

(continued)

Table 8-1 *(continued)*

Character Use	Character Sequence
Backspace	\b
Form feed	\f
New line	\n
Carriage return	\r
Horizontal tab	\t
Vertical tab	\v
Hexadecimal number, *N* (where *N* is the number of the character you want to display)	\x*N*
Octal number, *N* (where *N* is the number of the character you want to display)	*N*

Adding comments to your script

People tend to forget things. You might know how a script works on the day you create it and possibly even for a week after that. However, six months down the road, you may find that you don't remember much about the script at all. That's where comments come into play. Using comments helps you to remember what a script does, why it does it in a certain way, and even why you created the script in the first place. The following sections describe comments in more detail.

Using the % comment

Anytime MATLAB encounters a percent sign (%), it treats the rest of the line as a comment. *Comments* are simply text that is used either to describe what is happening in a script or to *comment out* lines of code that you don't want to execute. You can comment out lines of code during the troubleshooting process to determine whether a particular line is the cause of errors in your script. The "Analyzing Scripts for Errors" section, later in this chapter, provides additional details on troubleshooting techniques. Listing 8-3 shows how comments might appear in a script. You can also find this script in the Comments.m file supplied with the downloadable source code.

Listing 8-3: Using Comments to Make Code Easier to Read

```
% Tell MATLAB what to display on screen.
Prompt = ['Type your own name, but only if it isn''t ',...
          'Wednesday.\nType the name of the neighbor ',...
          'on your right on Wednesday.\nHowever, on ',...
```

```
                  'a Wednesday with a full moon, type the ',...
                  'name of\nthe neighbor on your left! '];

% Obtain the user's name so it can
% be displayed on screen.
Name = input(Prompt, 's');

% Output a message to make the user feel welcome.
disp(['Hello ', Name]);
```

Compare Listing 8-3 with Listing 8-2. You should see that the code is the same, but the comments make the code easier to understand. When you run this code, you see that the comments haven't changed how the script works. MATLAB also makes comments easy to see by displaying them in green letters.

Using the %% comment

MATLAB supports a double percent sign comment (%%) that supports special functionality in some cases. Here's how this comment works:

- ✔ Acts as a standard command in the Command window.
- ✔ Allows you to execute a portion (a section) of the code when using the Run and Advance feature.
- ✔ Creates special output when using the Publish feature.

The following sections describe the special %% functionality. You won't use this functionality all the time, but it's nice to know that it's there when you do need it.

Using Run and Advance

When you add a %% comment in the Editor window, MATLAB adds a section line above the comment (unless the comment appears at the top of the window), effectively dividing your code into discrete sections. To add a section comment, you type %%, a space, and the comment, as shown in Figure 8-5.

As with standard comments, the %% comment appears in green type. The line above the comment is your cue that this is a special comment. In addition, the position of the text cursor (the insertion point) selects a particular section. The selected section is highlighted in a pale yellow. Only the selected section executes when you click Run and Advance. Here's how sections work:

1. **Place the cursor at the end of the** Prompt = **line of code and then click Run and Advance.**

   Only the first section of code executes. Notice also that the text cursor comes to rest at the beginning of the second section.

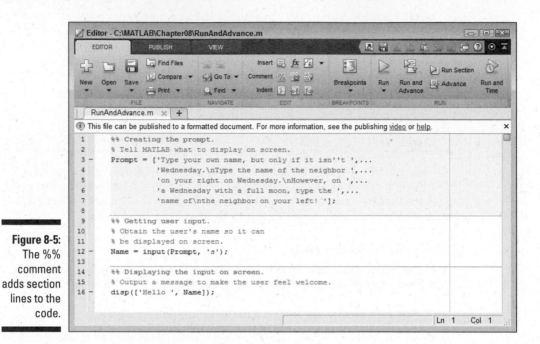

Figure 8-5:
The %%
comment
adds section
lines to the
code.

2. **Click Run and Advance.**

   The script displays a prompt asking for a name.

3. **Type a name and press Enter.**

   Only the second section of code executes. You don't see the script output.

4. **Place the cursor at the beginning of the second section and then click Run and Advance.**

   Steps 2 and 3 repeat themselves. You still don't see any script output.

5. **Click Run and Advance with the text cursor at the beginning of the third %% comment.**

   You see the script output (the correct output, in fact) without being asked for a name.

6. **Perform Step 5 as often as desired.**

   The application displays the script output every time without asking for any further information. Using this technique lets you execute just the part of a script that you need to test rather than run the entire script every time.

You can make small changes to the code and still run a particular section. For example, change `Hello` to `Goodbye` in the code shown previously in Figure 8-5. With the third section selected, click Run and Advance. The output displays a goodbye message, rather than a hello message, without any additional input.

Publishing information

The section comments let you easily document your script. This section provides just a brief overview of the publishing functionality, but it demonstrates just how amazing this feature really is. To start with, you really do need to create useful section comments — the kind that will make sense as part of a documentation package.

When creating the setup for the script you want to publish, you need to define the output format and a few essentials. The default format is HTML, which is just fine for this example. However, if you don't make one small change, the output isn't going to appear quite as you might like it to look. On the Publish tab of the Editor window, click the down arrow under Publish and choose Edit Publishing Options. You see the Edit Configurations dialog box, shown in Figure 8-6.

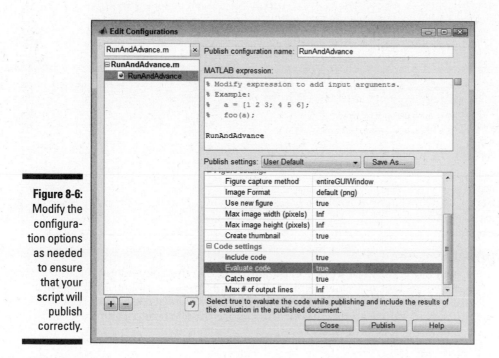

Figure 8-6: Modify the configuration options as needed to ensure that your script will publish correctly.

The Evaluate Code option evaluates your script and outputs the result as part of the documentation. Unfortunately, MATLAB can't evaluate input() functions as part of publishing the documentation for a script. As a consequence, you must set Evaluate Code to false. Click Publish. MATLAB produces an HTML page like the one shown in Figure 8-7.

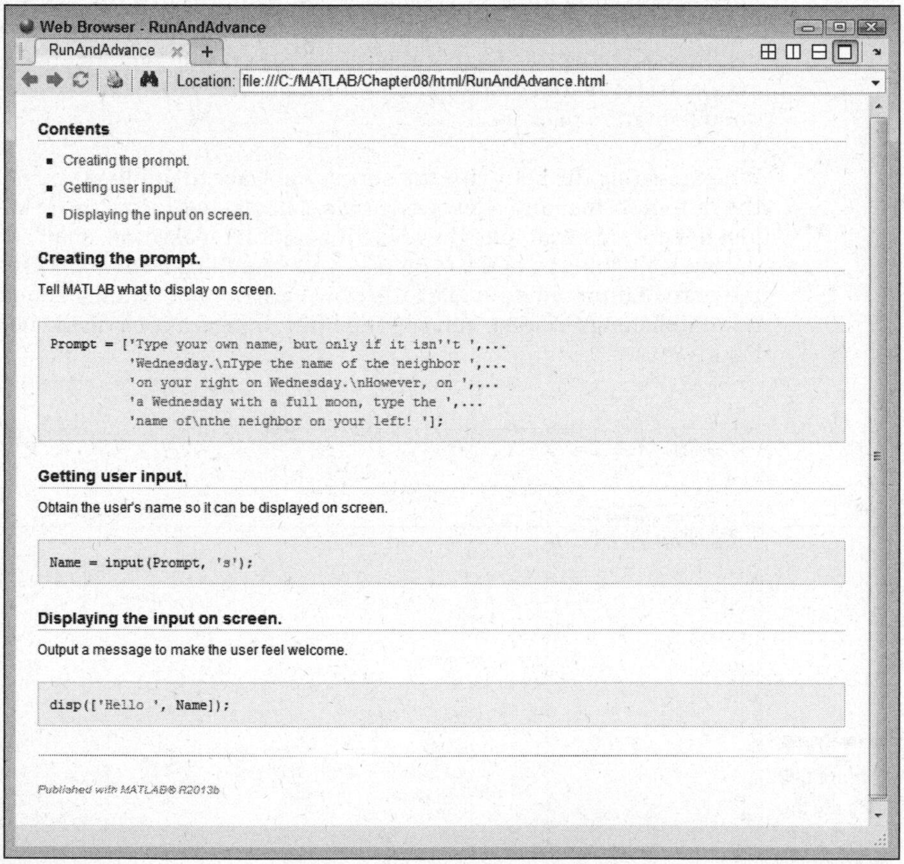

Figure 8-7:
The published documentation looks quite nice.

Considering the little work you put into creating the documentation, it really does look quite nice. In fact, it looks professional. When working with complex scripts, documentation like this really does serve a serious need. After you're done admiring your work, close the HMTL page and the Edit Configurations dialog box.

Revising Scripts

Scripts usually aren't perfect the first time you write them. In fact, editing them quite a few times is common. Even if the script does happen to attain perfection, eventually you want to add features, which means revising the script. The point is, you commonly see your scripts in the Editor window more than once. Here are some techniques you can use to open a script file for editing:

✔ Double-click the script's filename in the Current Folder window.

✔ Click the down arrow on the Open option of the Home tab of the MATLAB window and select the file from the list. (The list will contain every kind of file you have recently opened, not just script files.)

✔ Click the down arrow on the Open option of the Editor tab of the Editor window and select the file from the list. (The list will include only the most recently used script files.)

✔ Click Find Files in the Editor tab of the Editor window to display the Find Files dialog box. Enter a search criteria, such as *.m (where the asterisk is a wild-card character for all files) and click Find. Double-click the file you want to open in the resulting list.

✔ Locate the file using your platform's hard drive application (such as Windows Explorer in Windows or Finder on the Mac) and double-click the file entry.

It's a really bad idea to make changes to a script and then try to use it without testing it first. Always test your changes to ensure that they work as you intend them to. Otherwise, a change that you thought would work, could cause data damage or other problems.

Calling Scripts

Creating scripts without having some way to run them would be pointless. Fortunately, MATLAB lets you use scripts in all sorts of ways. The act of using a script — causing it to run — is known as *calling* the script. You can call scripts in these ways:

✔ Right-click the script file and select Run from the context menu that appears.

✔ Select the script file and press F9.

✔ Type the filename on the command line and press Enter. (Adding the extension isn't necessary.)

✔ Type the script filename in another script.

The last method of calling a script is the most important. It enables you to create small pieces of code (scripts) and call those scripts to create larger, more powerful, and more useful pieces of code. The next step is creating functions that can send information in and out of those smaller pieces of code. (You see the topic of functions explored in Chapter 9.)

Improving Script Performance

Scripts can run only so fast. The resources offered by your system (such as memory and processor cycles), the location of data, and even the dexterity of the user all come into play. Of course, with the emphasis on "instant" in today's society, faster is always better. With this in mind, the following list provides you with some ideas on how to improve your script performance. Don't worry if you don't completely understand all these bullets; you see most of these techniques demonstrated somewhere in the book. This list serves as a reference for when you're working on creating the fastest script possible:

✔ Create variables once instead of multiple times.

- Later in the book, you find a discussion on how to repeat tasks; creating variables inside these *loops* (bits of repeating code) is a bad idea.

- An application made up of smaller files might inadvertently re-create variables, so look for this problem as you analyze your application.

✔ Use variables to hold just one type of data. Changing the data type of a variable takes longer than simply creating a new one.

✔ Make code blocks as small as possible.

- Create several small script files rather than one large one.

- Define small functions rather than large ones.

- Simplify expressions and functions whenever possible.

✔ Use vectors whenever possible.

- Replace multiple scalar variables with one vector.

- Rely on vectors whenever possible to replace sparse matrices.

✔ Avoid running large processes in the background when using MATLAB.

Analyzing Scripts for Errors

Ridding an application of errors is nearly impossible. As complexity grows, the chances of finding absolutely every error diminishes. Everyone makes mistakes, even professional developers. So, it shouldn't surprise you that you might make mistakes from time to time as well. Of course, the important thing is to find the errors and fix them. The process of finding errors and fixing them is called *debugging*.

Sometimes the simplest techniques for finding errors is the best. Working with your script in sections is an important asset in finding errors. The "Using the %% comment" section, earlier in this chapter, describes how to create and use sections. When you suspect that a particular section has an error in it, you can run the code in that section multiple times as you look in the Workspace window to see the condition of variables that the code creates and the Command window to see the sort of output it creates.

Adding `disp()` statements to your code in various places lets you display the status of various objects. The information prints right in the Command window so that you can see how your application works over time. Removing the `disp()` statements that you've added for debugging purposes is essential after the session is over. You can do this by adding a % in front of the `disp()` statement. This technique is called *commenting out,* and you can use it for lines of code that you suspect might contain errors as well.

MATLAB also supports a feature called breakpoints. A *breakpoint* is a kind of stop sign in your code. It tells MATLAB to stop executing your code in a specific place so that you can see how the code is working. MATLAB supports two kinds of breakpoints:

- ✔ **Absolute:** The code stops executing every time it encounters the breakpoint. You use this kind of breakpoint when you initially start looking for errors and when you don't know what is causing the problem.

- ✔ **Conditional:** The code stops executing only when a condition is met. For example, a variable might contain a certain value that causes problems. You use this kind of breakpoint when you understand the problem but don't know precisely what is causing it.

To set a breakpoint, place the text cursor anywhere on the line and choose one of the options in the Breakpoints drop-down list on the Editor tab of the Editor window. When you set a breakpoint, a circle appears next to the line. The circle is red for absolute breakpoints and yellow for conditional breakpoints. Figure 8-8 shows both an absolute and a conditional breakpoint (you can't see the color in the printed version of the book, of course). Later chapters in the book demonstrate the use of breakpoints.

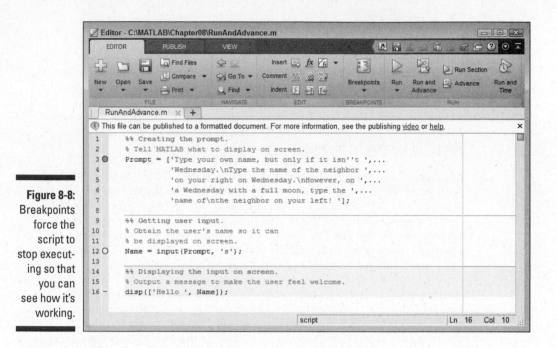

Figure 8-8:
Breakpoints force the script to stop executing so that you can see how it's working.

Creating error-handling code is also important in your application. Even though error handling doesn't fix an error, it makes the error less of a nuisance and can keep your application from damaging important data. Chapter 13 provides you with more ideas on how to locate and deal with errors in your script using error handling.

Chapter 9

Expanding MATLAB's Power with Functions

In This Chapter

▶ Finding and using built-in functions

▶ Defining and using your own functions

▶ Understanding other function types

Simplification is an important part of creating any useful application. The better you can outline what tasks the application performs in the simplest of terms, the easier it is to define how to interact with and expand the application. Understanding how an application works is the reason you use functions. A *function* is simply a kind of box in which you put code. The function accepts certain inputs and provides outputs that reflect the input received. It isn't important to understand precisely how the function performs its task unless your task is to modify that function, but being able to visualize what task the function performs helps you understand the application as a whole. The only requirement is that you understand the inputs and resulting outputs. In short, functions simplify the coding experience.

This chapter is about three sorts of functions. You have already used quite a few built-in functions, but simply using them may not be enough. You need to understand a little more about the inputs and outputs — the essentials of how the box works. On the other hand, you don't find out about the inner mechanisms of built-in functions in this chapter because you never need to know about those aspects.

You also get a chance to create your own functions in this chapter. The examples in previous chapters have been easy, so the need to provide simplification just isn't there. As the book progresses, you create more complex examples, so the need to simplify the code used in those examples becomes more important. Creating your own functions will make the examples easier to understand and your own code easier to understand as well.

MATLAB also supports some interesting alternatives to functions. They aren't functions in the traditional sense, but they make working with code simpler. These "special purpose" functions are used when the need arises to create code that is both efficient and elegant. The final part of this chapter provides a good overview of these special function types, and you see them used later in the book.

Working with Built-in Functions

Built-in functions are those that come with MATLAB or are part of an add-on product. You typically don't have source code for built-in functions and must treat them simply as black boxes. So far, you have relied exclusively on built-in functions to perform tasks in MATLAB. For example, when you use the `input()` and `disp()` functions in Chapter 8, you're using built-in functions. The following sections tell you more about built-in functions and how you can work with them in MATLAB to achieve specific objectives.

Learning about built-in functions

There are many ways you can learn about built-in functions, but if you already know the name of a function, one of the simplest makes use of the `help('`*function_name*`')` command, where *function_name* is the name of the function. Try it now. Type **help('input')** and press Enter in the Command window. You see output similar to the output shown in Figure 9-1.

MATLAB does provide some types of category help. For example, type **help('elfun')** and press Enter to see a listing of elementary math functions at your disposal. When you type **help('specfun')** and press Enter, you see a listing of specialized math functions.

Sometimes the help information provided by the `help()` function becomes excessively long. In this case, you can use the `more()` function to present the information a page at a time. Before you use the `help()` function, type **more('on')** and press Enter to put MATLAB in paged mode. When the help information is more than a page in length, you see a

```
--more--
```

prompt at the bottom of the screen. Press the spacebar to see the next page. If you want to see only the next line, press Enter instead. When you finish reviewing help, type **more('off')** and press Enter to turn off paged mode.

```
Command Window                                              ▭ ▭ ✕

New to MATLAB? Watch this Video, see Examples, or read Getting Started.    ✕

>> help('input')
 input  Prompt for user input.
     NUM = input(PROMPT) displays the PROMPT string on the screen, waits for
     input from the keyboard, evaluates any expressions in the input, and
     returns the value in NUM. To evaluate expressions, input accesses
     variables in the current workspace. If you press the return key without
     entering anything, input returns an empty matrix.

     STR = input(PROMPT,'s') returns the entered text as a MATLAB string,
     without evaluating expressions.

     To create a prompt that spans several lines, use '\n' to indicate each
     new line. To include a backslash ('\') in the prompt, use '\\'.

     Example:

         reply = input('Do you want more? Y/N [Y]:','s');
         if isempty(reply)
             reply = 'Y';
         end

     See also keyboard.

     Reference page in Help browser
        doc input

fx >>
```

Figure 9-1:
Obtain help
directly from
MATLAB
for built-in
functions
you know.

Although the help() function is really useful because it displays the information you need directly in the Command window, sometimes the doc() function is a better choice. When using the doc() function, you see a nicely formatted output that includes links to example code and other information. Type **doc('input')** and press Enter, and you see the output shown in Figure 9-2. This is the option you should use when you want to get an in-depth view of a function rather than simply jog your memory as part of writing an application. In addition, when you find that the help() function is less helpful than you'd like, the doc() function generally provides more information.

Using help() may not always be possible because you don't know the precise name of whatever you need to find. Another useful function is docsearch(). You use this function when you have some idea, but not a precise one, of what you need to find. For example, type **docsearch('input')** and press Enter in the Command window. This time you see a list of potential entries to query, as shown in Figure 9-3. Notice that the input() function is still the first entry on the list, but you have a number of other choices as well.

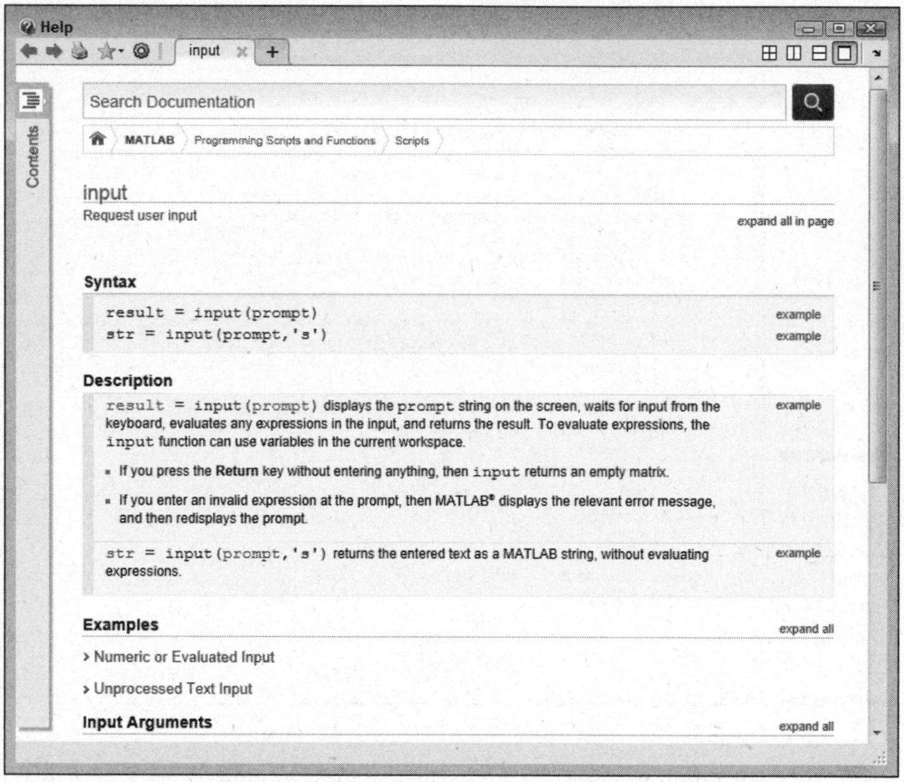

Figure 9-2:
Use the
`doc()`
function
when
you need
in-depth
informa-
tion about
a built-in
function.

One of the more interesting ways to search for built-in functions is to use the `lookfor()` function. In this case, MATLAB doesn't look in the documentation; rather, it looks in the source code files. This kind of search is important because you can sometimes see connections between functions this way and find alternatives that might not normally occur to you. To see how this kind of search works, type **lookfor('input')** and press Enter. You see the output shown in Figure 9-4. Notice that the `input()` function is in the list, but it doesn't appear at the top because the search doesn't sort the output by likely candidate.

If you really want to know more about the built-in functions from a coding perspective, start with the `which()` function, which tells you the location of the built-in function. For example, type **which('input')** and press Enter. You see the location of this built-in function on your system. On my system, I receive this output: `built-in (C:\Program Files\MATLAB\R2013b\ toolbox\matlab\lang\input)`.

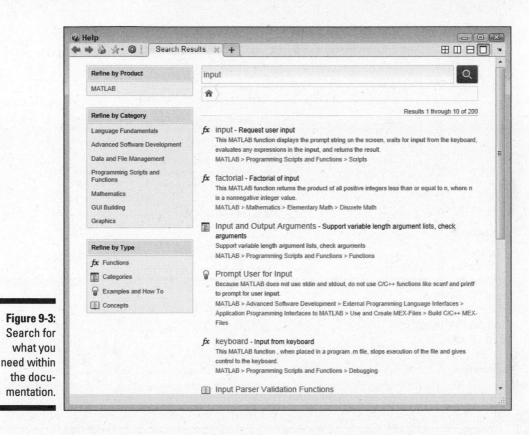

Figure 9-3:
Search for
what you
need within
the docu-
mentation.

At this point, you know that input() is found in the lang folder. However, you really don't know what related functions might be in the same folder. Use the what() function to locate additional information about the content of the lang folder. To see this for yourself, type **what('lang')** and press Enter. You see the output shown in Figure 9-5. Notice that the output includes the disp() function that you used with the input() function in Chapter 8. However, you also see a number of other interesting functions in the list that could prove useful.

Not shown in Figure 9-5 is a listing of classes and packages found in the lang folder. Classes and packages are simply two other ways of packaging functionality within MATLAB. However, these two packaging methods provide more functionality than functions do in most cases, so it pays to look them up to see what sorts of things you can do with them. Using the doc() and help() functions provides you with information about the classes and packages.

```
Command Window                                                    □ □ ✕
① New to MATLAB? Watch this Video, see Examples, or read Getting Started.    ✕
>> lookfor('input')                                                    ⊙ ▲
inputParser              - Construct input parser object
axescheck                - Process leading Axes object from input lis
datetickstr              - Returns the date string associated with th
ginput                   - Graphical input from mouse.
inputdlg                 - Input dialog box.
menu                     - Generate a menu of choices for user input.
datatipinfo              - Produce short description of input variabl
deal                     - Deal inputs to outputs.
isa                      - Determine if input is object of specified
superiorfloat            - return 'double' or 'single' based on the s
blkdiag                  - Block diagonal concatenation of matrix inp
iscolumn                 - True if input is a column vector.
ismatrix                 - True if input is a matrix.
isrow                    - True if input is a row vector.
isscalar                 - True if input is a scalar.
isvector                 - True if input is a vector.
input                    - Prompt for user input.
inputname                - Input argument name.
iskeyword                - Check if input is a keyword.
nargchk                  - Validate number of input arguments.
nargin                   - Number of function input arguments.
narginchk                - Validate number of input arguments.
varargin                 - Variable length input argument list.
automesh                 - returns true if the inputs should be passe
wavrecord                - Record sound using Windows audio input dev
fx >>
◄                          III                                        ►
```

Figure 9-4:
In some cases, you need to look for associations as part of your search.

Using online information sources

Although you can obtain a lot of help using just the functionality that MATLAB provides, it also pays to look online for help as needed. The best place to look for help on built-in functions is the MATLAB site at http://www.mathworks.com/help/matlab/functionlist.html. This site provides a complete list of built-in functions in category order. If you want an alphabetical list of functions, try the site at http://man.fsid.cvut.cz/matlab6_r13/techdoc/ref/refbookl.html. The Internet is packed with all sorts of useful information about MATLAB functions.

Be sure to exercise caution when using online sources. If possible, check for a date or version number for the information. For example, the site at http://www.eng.umd.edu/~austin/ence202.d/matlab-functions.html looks interesting at first, but then you see that the information was current as of 1984, so it's a dubious source of information at best. Information tends to live on the Internet nearly forever, so always verify that the information you're using is current.

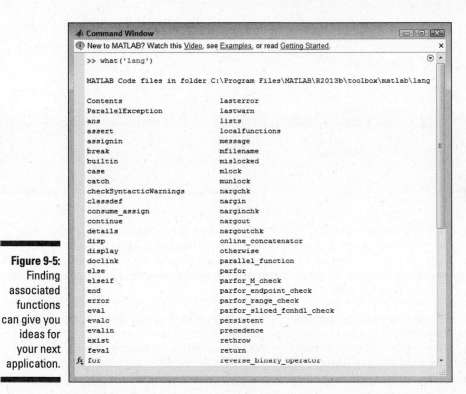

Figure 9-5:
Finding
associated
functions
can give you
ideas for
your next
application.

Sending data in and getting data out

The essence of a function is that it presents you with a black box. In most cases, you send data in, it whirls around a bit, and then data comes back out. Managing data is an essential part of most functions.

Of course, some functions require only input, some provide only output, and some perform tasks other than work directly with data. For example, the `clc()` clears the Command window and doesn't require any data input or produce any data output to perform the task. Every function does something; creating one that does nothing would be pointless.

The problem for many people is determining the input and output requirements for the built-in functions. The best way to discover this information is to use the `help()` or `doc()` functions. The `doc()` function is actually the easiest to use in this case. The input and output arguments appear at the bottom of the help screen. To see this for yourself, type **doc('input')** and press Enter. Scroll down to the bottom of the resulting page and you see the inputs and outputs shown in Figure 9-6.

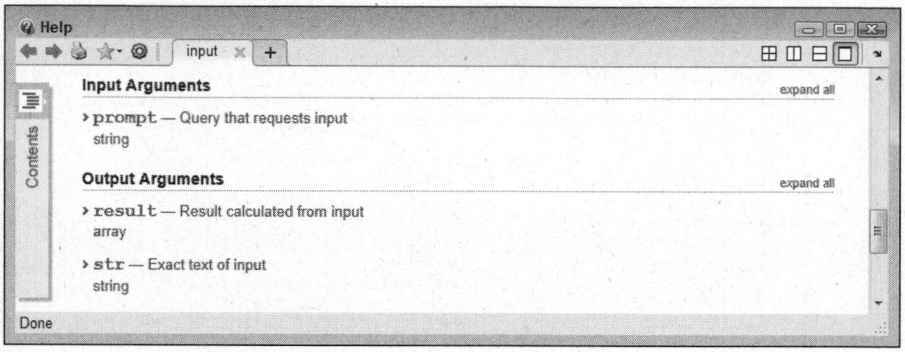

Figure 9-6:
The doc()
function
lists inputs
and outputs
in an easily
found form.

Input Arguments   expand all

> **prompt** — Query that requests input
  string

Output Arguments   expand all

> **result** — Result calculated from input
  array

> **str** — Exact text of input
  string

In this case, you see that the input argument is a prompt and that you must provide this input as a string. The documentation explains that the prompt is there to ask the user for a specific kind of input. The output can take two forms: an array that is calculated from the input or a string that contains the precise text the user has typed.

When you see a dual output for a function, it means that you need to tell the function what sort of output to provide or that there is a default. In this case, the input() function requires that you supply a second argument, 's', to obtain the string output. The default is to provide the calculated array.

Creating a Function

Functions represent another method for packaging your code. They work as an addition to scripts rather than a replacement for them. Scripts and functions each have a particular place to occupy in your MATLAB toolbox. The first section that follows explains these differences and helps you understand when you would use a script or a function. In some cases, it doesn't matter too much, but in other cases the wrong choice can cause you a lot of frustration and wasted time.

The remainder of the sections that follow help you create custom functions of various types. You start with a simple function that doesn't require any input or output to perform a task. After that, you start to build functions with greater complexity that are also more flexible because they do accept input and produce output. Functions can be as simple or complex as needed to perform a task, but simpler is always better (an emphasis of this chapter as a whole).

Understanding script and function differences

A *script* is a method of packaging a procedure — in other words, a series of steps that you use to perform a task. Some people have compared scripts to keyboard macros or other forms of simple step recording. On the other hand, a *function* is a method of packaging a transformation — code that is used to manage data in some manner or to perform a task that requires better data handling than a script can provide. Both types of packages contain code of a sort, but each packaging method is used differently.

Scripts and functions also handle data differently. A script makes all the variables that it contains part of the workspace. As a result, after the script runs you can easily see all the variables that the script contains as well as their ending values. A function hides its variables, and the variables become unavailable after the function runs. As a result, the actual data that the function uses internally isn't visible, and you must supply any required inputs every time you run the function.

As you see later in this section, a function also has a special header that identifies the function name, the inputs that it requires, and the outputs it provides. A function is a formal sort of coding method that's more familiar to developers. However, functions also provide greater flexibility because you can control the environment in which they perform tasks with greater ease.

The use of inputs and outputs reduces the potential for contamination by data left over from a previous run and, like Las Vegas, what happens in the function stays in the function. This feature is a big advantage: You can use the same name in a function as you would outside it without interference, and doing so avoids a lot of confusion.

Both scripts and functions reside in files that have an .m extension. The immediately noticeable difference between the two is that a script lacks a header. Functions always have the header that you see in the "Writing your first function" section, later in this chapter.

Understanding built-in function and custom function differences

Built-in functions (those provided with MATLAB) and custom functions (those you create yourself or that come as part of a third party product) differ in at least one important aspect. The custom functions come with source code. You can modify this source code as needed to meet your particular needs.

The built-in `input()` function comes with MATLAB, and you can find it in the `input.m` file in the `toolbox\matlab\lang` directory used to contain part of the files for your MATLAB installation. However, if you open that file, you see documentation but no source code. The source code is truly part of MATLAB, and you can't edit it. You can modify the documentation as necessary with your own notes, but this really isn't a recommended procedure because the next MATLAB update will almost certainly overwrite your changes.

Writing your first function

Creating a function is only slightly more work than creating a script. In fact, the two processes use the same editor, so you're already familiar with what the editor can provide in the way of help. The various Editor features you'd use for creating a script all work the same way with functions, too. (You have access to the same double percent sign (%%) for use with sections, for example.) The following steps get you started creating your first function. You can also find this function in the `SayHello.m` file supplied with the downloadable source code.

1. **Click the arrow under the New entry on the Home tab of the MATLAB menu and select Function from the list that appears.**

   You see the Editor window shown in Figure 9-7. Notice that the editor already has a function header in place for you, along with the inputs, outputs, and documentation comments.

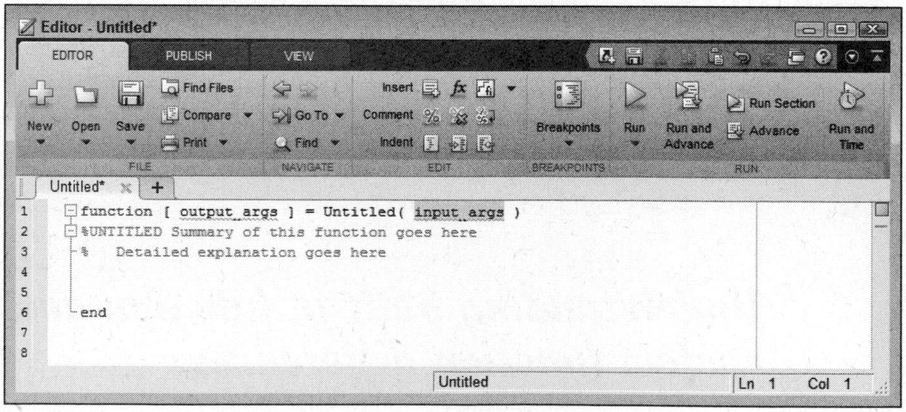

Figure 9-7: The Editor window helps you create new functions.

Figure 9-7 may look a little complex, but that's because MATLAB includes a number of optional elements that you will see in action later in the chapter. A function has three requirements:

- A function always begins with the word `function`.

- You must include a function name.

- A function must always end with the keyword `end`.

2. **Delete** `output_args`.

Functions aren't required to have output arguments. In order to keep things simple for your first function, you're not going to require any inputs or outputs

An *argument* is simply a word for an individual data element. If you supply a number to a function, the number is considered an argument. Likewise, when you supply a string, the entire string is considered just one argument. A vector, even though it contains multiple numbers, is considered a single argument. Any single scalar or object that you provide as input or that is output from the function is considered an argument.

3. **Delete** `input_args`.

Functions aren't required to have input arguments.

4. **Change the function name from** `Untitled` **to** `SayHello`.

Your function should have a unique name that reflects its purpose. Avoiding existing function names is essential. Before you name your function, test the name you're considering by typing **help('*NameOfYourFunction*')** and pressing Enter. If the function already exists, you see a help screen. Otherwise, MATLAB denies all knowledge of the function, and you can use the function name you have chosen.

Always provide help information with the functions you create. Otherwise, the `help()` function won't display any help information and someone could think that your function doesn't exist. If you want to be absolutely certain that there is no potential conflict between a function you want to create and an existing function (even a poorly designed one), use the `exist()` function instead, such as `exist('SayHello')`. When the function exists, you see an output value of 2. Otherwise, you see an output value of 0.

5. **Change the comments to read like this:**

```
%SayHello()
%    This function says Hello to everyone!
```

Notice that the second line is indented. The indentation tells MATLAB that the first line is a title and the second is text that goes with the title. Formatting your comments becomes important when working with functions. Otherwise, you won't see the proper help information when you request it.

6. **Add the following code after the comment:**

```
disp('Hello There!');
```

The function simply displays a message onscreen.

7. **Click Save.**

You see the Select File for Save As dialog box, shown in Figure 9-8.

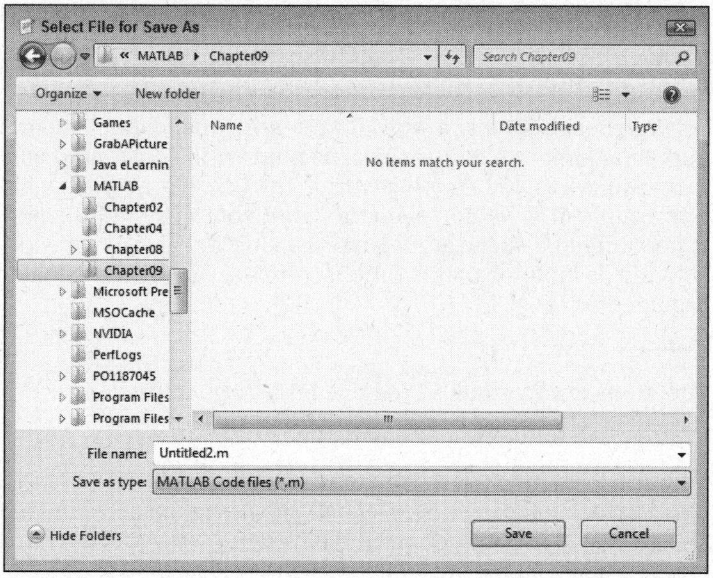

Figure 9-8:
You must
save your
function to
disk in order
to use it.

8. **Select the Chapter09 directory for the source code for this book, type** SayHello,m **in the File Name field and then click Save.**

MATLAB saves your function as SayHello.m.

The filename you use to store your function must match the name of the function. MATLAB uses the filename to access the function, not the function name that appears in the file. When there is a mismatch between the function name and the filename, MATLAB displays an error message.

Using the new function

You have a shiny new function and you're just itching to use it. Before you can use the function, you must make sure that the directory containing the function file is part of the MATLAB path. You can achieve this goal in two ways:

✔ Double-click the directory entry in the Current Folder window.

✔ Right-click the directory entry in the Current Folder window and choose Add to Path⇨Selected Folders and Subfolders from the context menu.

You can try your new function in a number of ways. The following lists contains the most common methods:

✔ Click Run in the Editor window, and you see the output in the Command window. However, there is a little twist with functions that you discover in the upcoming "Passing data in" section of the chapter. You can't always click Run and get a successful outcome, even though the function will always run.

✔ Click Run and Advance in the Editor window. (This option runs the selected section when you have sections defined in your file.)

✔ Click Run and Time in the Editor window. (This option outputs profiling information — statistics about how the function performs — for the function.)

✔ Type the function name in the Command window and press Enter.

Your function also has help available with it. Type **help('SayHello')** and press Enter. MATLAB displays the following help information:

```
SayHello()
    This function says Hello to everyone!
```

The output is precisely the same as it appears in the function file. The doc() function also works. Type **doc('SayHello')** and press Enter. You see the output shown in Figure 9-9. Notice how the title is presented in a different color and font than the text that follows.

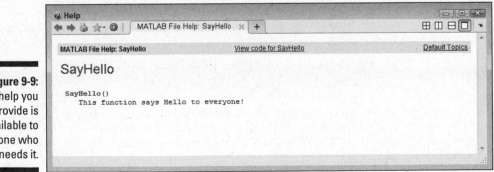

Figure 9-9:
The help you provide is available to anyone who needs it.

Passing data in

The SayHello() function is a little limited. For one thing, it can't greet anyone personally. To make SayHello() a little more flexible, you need to pass some information to it in the form of an input argument. The following steps help you create an updated SayHello() that accepts input arguments. You can also find this function in the SayHello2.m file supplied with the downloadable source code.

1. **Click the down arrow under the Save option on the Editor tab of the Editor window and choose Save As.**

    You see the Select File for Save As dialog box. (Refer to Figure 9-8.)

2. **Type** SayHello2.m **in the File Name field and click Save.**

    MATLAB saves the function that you created earlier using a new name. Notice that the function name is now highlighted in orange. The highlight tells you that the function name no longer matches the filename.

3. **Change the function name from** SayHello **to** SayHello2.

    The orange highlight disappears when you place the text cursor in another location in the Editor window.

4. **Add the input argument** Name **to the function header so that the header looks like this:**

    ```
    function [ ] = SayHello2( Name )
    ```

    Notice that Name is now highlighted in orange to show that you haven't used it anywhere. The highlight will go away after you make the next change. The Editor window always displays an orange highlight when it detects a problem with your code. It's important to realize that the Code Analyzer feature of MATLAB detects only potential errors. (You can read more about the Code Analyzer at http://www.mathworks.com/help/matlab/matlab_prog/matlab-code-analyzer-report.html.) It can't absolutely tell you that a problem exists, so you need to look carefully at the highlights.

5. **Change the** disp() **function call so that it looks like this:**

    ```
    disp(['Hello There ', Name, '!']);
    ```

    The disp() function now requires use of the concatenation operator that was introduced in Chapter 8 to combine the text with the input argument. The output will contain a more personalized message.

6. **Click Run.**

    MATLAB displays a message telling you that the SayHello2() function requires input, as shown in Figure 9-10. You see this message every time you try to use Run to start the function because functions don't store argument information as scripts do.

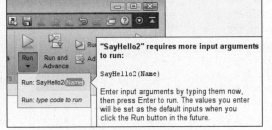

Figure 9-10: MATLAB knows to ask you for the input arguments as needed.

7. **Type** `'Ann'` **and press Enter.**

   You see the expected output of:

   ```
   Hello There Ann!
   ```

8. **Type** `SayHello2('Josh')` **in the Command window and press Enter.**

   You see the expected output.

Passing data out

When functions manipulate data, they pass the result back to the caller. The *caller* is the code that called the function. The following steps help you create an updated `SayHello2()` function that passes back the string it creates. You can also find this function in the `SayHello3.m` file supplied with the downloadable source code.

1. **Click the down arrow under the Save option on the Editor tab of the Editor window and choose Save As.**

   You see the Select File for Save As dialog box. (Refer to Figure 9-8.)

2. **Type SayHello3.m in the File Name field and click Save.**

   MATLAB saves the function you created earlier using a new name.

3. **Change the function name from** `SayHello2` **to** `SayHello3`.

   The orange highlight disappears when you place the text cursor in another location in the Editor window.

4. **Type HelloString in the square brackets before the function name so that your function header looks like this:**

   ```
   function [ HelloString ] = SayHello3( Name )
   ```

Using optional arguments

Dealing with optional arguments requires some MATLAB features that are covered in later chapters, so you can feel free to skip this sidebar for now and come back to it later. The current version of the SayHello2() function requires an argument. You must supply a name or the function won't run. A number of functions that you have already used, such as disp(), provide optional arguments — you can choose to use them or not. Optional arguments are important because there are times when a default argument works just fine. In fact, you can change SayHello2() so that it doesn't require an argument.

To make SayHello2() work without an input argument, you must assign a value to Name, but only if Name doesn't already have a value. In order to make the assignment, you first need to determine whether Name does have a value by checking a variable that MATLAB provides for what you called nargin (for *N*umber of *A*rguments *I*nput). If nargin equals 1, then the caller — the code that called this function — has provided an input. Otherwise, you need to supply the input to Name. Here's the updated version of SayHello2().

```
function [ ] = SayHello2(
    Name )
%SayHello()
%   This function says Hello
    to everyone!
if nargin < 1
    Name = 'Good Looking';
end

disp(['Hello There ', Name,
    '!']);

end
```

The additional code states that if nargin is less than 1, the function needs to assign it a value of 'Good Looking'. Otherwise, the function uses the name provided by the caller. To test this code, type **SayHello2** and press Enter in the Command window. You see Hello There Good Looking! as output. Of course, the function could be broken, so type **SayHello2('Selma')** and press Enter. You see Hello There Selma! as output, so the function works precisely as it should and Name is now an optional argument.

The function now returns a value to the caller. You see the orange highlight again because SayHello3() doesn't assign a value to HelloString yet.

5. Modify the function code to include an assignment to HelloString, like this:

```
HelloString = ['Hello There ', Name, '!'];
disp(HelloString);
```

The function now assigns a value to HelloString and then uses that value as output. It also returns the output to the caller.

6. Save the changes you've made.

7. **Type** Output = SayHello3('Ambrose') **in the Command window and press Enter.**

   You see the following output:

   ```
   Hello There Ambrose!
   Output =
   Hello There Ambrose!
   ```

8. **Type** disp(Output) **in the Command Window and press Enter.**

   You see the expected greeting as output.

Creating and using global variables

Functions normally use *local variables* — that is, they aren't visible to anyone but the function. Using local variables avoids confusion because each function is self-contained. In addition, using local variables makes functions more secure and reliable because only the function can access the data in the variable.

You may find that you need to make a variable visible, either because it is used by a number of functions or the caller needs to know the value of the variable. When a function makes a local variable visible to everyone, it becomes a *global variable*. Global variables can be misused because they're common to every function that wants to access them and they can present security issues because the data becomes public.

The following steps show how to create a global variable. You can also find these functions in the SayHello4.m and SayHello5.m files supplied with the downloadable source code.

1. **Click the down arrow under the Save option on the Editor tab of the Editor window and choose Save As.**

   You see the Select File for Save As dialog box. (Refer to Figure 9-8.)

2. **Type** SayHello4.m **in the File Name field and click Save.**

   MATLAB saves the function you created earlier using a new name.

3. **Change the function name from** SayHello3 **to** SayHello4.

   The orange highlight disappears when you place the text cursor in another location in the Editor window.

4. **Remove** HelloString **from the square brackets before the function name so that your function header looks like this:**

   ```
   function [ ] = SayHello4( Name )
   ```

When a variable is global, you can't return it as data from a function call. The data is already available globally, so there is no point in returning it from the function.

5. **Change the** `HelloString` **assignment so that it now contains the global keyword, as shown here:**

```
global HelloString;
HelloString = ['Hello There ', Name, '!'];
```

6. **Save the changes you've made.**

7. **Type** SayHello4('George') **in the Command window and press Enter.**

You see the following output:

```
Hello There George!
```

At this point, there is a global variable named `HelloString` sitting in memory. Unfortunately you can't see it, so you don't really know that it exists for certain.

8. **Perform Steps 1 through 3 to create** `SayHello5()`.

9. **Modify the** `SayHello5()` **code so that it looks like this:**

```
function [ ] = SayHello5( )
%SayHello()
%   This function says Hello to everyone!
global HelloString
disp(HelloString);

end
```

Notice that `SayHello5()` doesn't accept input or provide output arguments. In addition, it only declares `HelloString`; it doesn't actually assign a value to it, so the function should fail when it calls the `disp()` function.

10. **Type** SayHello5 **in the Command window and press Enter.**

You see Hello There George! as the output. The global variable really is accessible from another function.

Using subfunctions

A single function file can contain multiple functions. However, only one function, the *primary function* (the one that has the same name as the original file), is callable. Any other functions in the file, known as *subfunctions,* are local to that file. The primary function or other subfunctions can call on any subfunction, as long as that subfunction appears in the same file.

The main reason to use subfunctions is to simplify your code by breaking it into smaller pieces. In addition, placing common code in a subfunction means that you don't have to copy and paste it all over the place — you have to write it only once. As far as anyone else is concerned, however, the file contains only one function. The inner workings of your code are visible only to you and anyone else who can view the source code.

Listing 9-1 shows an example of how a subfunction might work. You can also find this function in the `SayHello6.m` file supplied with the downloadable source code.

Listing 9-1: Creating a Subfunction

```
function [ HelloString ] = SayHello6 ( Name )
%SayHello()
%   This function says Hello to everyone!
HelloString = [GetGreeting(), Name, '!'];
disp(HelloString);

end

function [ Greeting ] = GetGreeting ( )
Greeting = 'Hello There ';
End
```

This code is actually another version of the SayHello3 code that you worked with earlier. The only difference is that the greeting is now part of the `GetGreeting()` subfunction, rather than a simple string. Notice that `SayHello6()` can call `GetGreeting()`, using the same technique that it could use for any other function.

After you create this code, type **Output = SayHello6('Stan')** in the Command window and press Enter. You see the following output:

```
Hello There Stan!
Output =
Hello There Stan!
```

The output is precisely as you expect. However, now type **GetGreeting()** and press Enter in the Command window. Instead of a greeting, you see an error message:

```
Undefined function or variable 'GetGreeting'.
```

The `GetGreeting()` subfunction isn't accessible to the outside world. As a result, you can use `GetGreeting()` with `SayHello6()` and not have to worry about outsiders using the subfunction incorrectly.

Nesting functions

You can also nest functions one inside the other in MATLAB. The *nested function* physically resides within the primary function. The difference between a primary function and a nested one is that the nested function can access all the primary function data, but the primary function can't access any of the nested function data.

In all other respects, subfunctions and nested functions behave in a similar manner (for example, you can't call either subfunctions or nested functions directly). Listing 9-2 shows a typical example of a nested function. You can also find this function in the SayHello7.m file supplied with the downloadable source code.

Listing 9-2:   Creating a Nested Function

```
function [ HelloString ] = SayHello7( Name )
%SayHello()
%   This function says Hello to everyone!
HelloString = [GetGreeting(), Name, '!'];
disp(HelloString);

    function [ Greeting ] = GetGreeting ( )
    Greeting = 'Hello There ';
    end
end
```

This is another permutation of the SayHello3() example, but notice how the GetGreeting() nested function now resides inside SayHello7(). After you create this code, type **Output = SayHello7('Stan')** in the Command window and press Enter. You see the following output:

```
Hello There Stan!
Output =
Hello There Stan!
```

Using Other Types of Functions

MATLAB supports a few interesting additions to the standard functions. In general, these additions are used to support complex applications that require unusual programming techniques. However, it pays to know that the functions exist for situations in which they come in handy. The following sections provide a brief overview of these additions.

Inline functions

An *inline function* is one that performs a small task and doesn't actually reside in a function file. You can create an inline function right in the Command window if you want. The main purpose for an inline function is to make it easier to perform a calculation or manipulate data in other ways. You use an inline function as a kind of macro. Instead of typing a lot of information every time, you define the inline function once and then use the inline function to perform all the extra typing.

To see an inline function in action, type **SayHello8 = inline('["Hello There ", Name, "!"]')** in the Command window and press Enter. You see the following output:

```
SayHello8 =
     Inline function:
     SayHello8(Name) = ['Hello There ', Name, '!']
```

This function returns a combined greeting string. All you need to do is type the function name and supply the required input value. Test this inline function by typing **disp(SayHello8('Robert'))** and pressing Enter. You see the expected output:

```
Hello There Robert!
```

Notice that the inline function doesn't actually include the disp() function call. An inline function must return a value, not perform output. If you try to include the disp() function call, you see the following error message:

```
Error using inlineeval (line 15)
Error in inline expression ==> disp(['Hello There ',
          Name, '!'])
 Too many output arguments.
Error in inline/subsref (line 24)
              INLINE_OUT_ = inlineeval(INLINE_INPUTS_,
          INLINE_OBJ_.inputExpr, INLINE_OBJ_.expr);
```

Anonymous functions

An anonymous function is an even shorter version of the inline function. It can contain only a single executable statement. The single statement can accept input arguments and provide output data.

To see how an anonymous function works, type **SayHello9 = @(Name) ['Hello There ', Name, '! ']** and press Enter. You see the following output:

```
SayHello9 =
    @(Name) ['Hello There ',Name,'!']
```

The at (@) symbol identifies the code that follows as an anonymous function. Any input arguments you want to accept must appear in the parentheses that follow the @ symbol. The code follows after the input argument declaration. In this case, you get yet another greeting as output.

To test this example, type **disp(SayHello9('Evan'))** in the Command window and press Enter. You see the following output:

```
Hello There Evan!
```

You generally use anonymous functions for incredibly short pieces of code that you need to use repetitively. Inline functions execute more slowly than anonymous functions for a comparable piece of code. So whenever possible, use an anonymous function in place of an inline function. However, inline functions also provide the extra flexibility of allowing multiple lines of code, so you need to base your decision partly on how small you can make the code that you need to execute.

Chapter 10

Adding Structure to Your Scripts

· ·

In This Chapter

▶ Adding decision-making to your scripts

▶ Using recursive functions

▶ Repeating tasks

▶ Defining menus

· ·

*T*he scripts and functions you have created so far have all performed a series of tasks, in order, one at a time. However, sometimes it's important to skip steps or to perform the same step more than once. Humans make decisions about what to do and how often to do it with ease, but computers need a little help. You must provide specific instructions as to when a decision is required and when to do something more than once.

As part of discovering how decisions are made and just how repetition works, you see two practical examples of how to employ your new skills. The first is a technique called *recursion,* which is simply a method of performing a task more than once, but in an elegant way. Using recursion makes solving some math problems significantly easier. The second involves the use of a menu. Most multifunction applications rely on menus to allow a user to select one option out of a number of possibilities.

Making Decisions

When you come to an intersection, you make a decision: go through or stop. If the light is red, you stop. However, when the light is green, you go through the intersection. The "if this condition exists, then do this" structure is something humans use almost constantly. In fact, we sometimes don't think about it consciously at all. A decision is often made without conscious thought because we have made it so many times.

A computer needs guidance in order to make a decision. It lacks a subconscious and therefore needs to work through each decision. When working with MATLAB, you might need to tell the computer that if a value is above a certain level, the computer should perform a certain set of steps. Perhaps you need to compensate in the computation of a result (some factor has affected the result beyond the normal value) or you simply need to look at the formula in a different way. The decision could be procedural — a user selects a particular option and the computer needs to perform the associated task. The thing to remember about a computer is that it follows whatever you tell it precisely, so you need to provide precise decision-making instructions.

The following sections describe two different decision-making structures that MATLAB provides: the if statement and the switch statement. Each of these statements has a specific format as well as specific times when you'd use it.

Decision-making code has a number of terms associated with it. A *statement* simply indicates what the code should do. It's the line of code that appears first in a block of tasks. A *structure* describes the statement and all the code that follows until the end keyword is reached.

Using the if statement

The simplest decision to make is whether to do something — or not. However, you might need to decide between two alternatives. When a situation is true, you perform one task, but when it's false, you perform another task. There are still other times when you have multiple alternatives and must choose a course of action based on multiple scenarios using multiple related decisions. The following sections cover all these options.

Making a simple decision

Starting simply is always best. The if statement makes it possible to either do something when the condition you provide is true or not do something when the condition you provide is false. The following steps show how to create a function that includes an if statement. You can also find this function in the SimpleIf.m file supplied with the downloadable source code.

1. **Click the arrow under the New entry on the Home tab of the MATLAB menu and select Function from the list that appears.**

   You see the Editor window, shown in Figure 10-1.

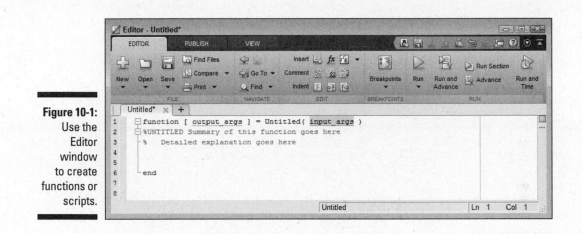

Figure 10-1:
Use the
Editor
window
to create
functions or
scripts.

2. **Delete** output_args.

   The example doesn't provide an output argument, but it does require an input argument.

3. **Change the function name from** Untitled **to** SimpleIf.

   The primary function name must match the name of the file.

4. **Change** input_args **to** Value.

   The term input_args is used only to tell you that you need to provide input arguments. In this case, the function receives a value from the caller to use in the decision-making process.

5. **Type the following code into the function between the comment and the** end **keyword.**

   ```
   if Value > 5
       disp('The input value is greater than 5!');
   end
   ```

   This code makes a simple comparison. When the input argument, Value, is greater than 5, the function tells you about it. Otherwise, the function doesn't provide any output at all.

6. **Click Save.**

   You see the Select File for Save As dialog box, shown in Figure 10-2. Notice that the File Name field has the correct filename entered for you. This is the advantage of changing the function name before you save the file for the first time.

7. **Click Save.**

   The function file is saved to disk.

8. **Type** SimpleIf(6) **and press Enter in the Command window.**

   You see the following output:

   ```
   The input value is greater than 5!
   ```

9. **Type** SimpleIf(4) **and press Enter in the Command window.**

   The function doesn't provide any output. Of course, this is the expected reaction.

Adding an alternative option

Many decisions that people make are choices between two options. For example, you might go to the beach today, or choose to stay home and play dominoes based on whether it is sunny. When the weather is sunny, you go to the beach. MATLAB has a similar structure. The application chooses between two options based on a condition. The second option is separated from the first by an else clause — the application performs the first task, or else it performs the second. The following steps demonstrate how the else clause works. (These steps assume that you completed the SimpleIf example in the preceding section.) You can also find this function in the IfElse.m file supplied with the downloadable source code.

1. **In the Editor window, with the** `SimpleIf.m` **file selected, click the down arrow under Save and choose Save As from the list that appears.**

   You see the Select File for Save As dialog box. (Refer to Figure 10-2.)

2. **Type** IfElse.m **in the File Name field and click Save.**

   MATLAB saves the example using a new name.

3. **Replace the** `SimpleIf` **function name with** `IfElse`.

4. **Add the following code after the** `disp()` **function call:**

   ```
   else
        disp('The input value is less than 6!');
   ```

   The function can now respond even when the primary condition isn't met. When `Value` is greater than 5, you see one message; otherwise, you see the other message.

5. **Click Save.**

   The function file is saved to disk.

6. **Type** IfElse(6) **and press Enter in the Command window.**

   You see the following output:

   ```
   The input value is greater than 5!
   ```

7. **Type** IfElse(4) **and press Enter in the Command window.**

   You see the following output:

   ```
   The input value is less than 6!
   ```

   The example demonstrates that you can provide alternative outputs depending on what is happening within the application. Many situations arise in which you must choose an either/or type of condition.

Creating multiple alternative options

Many life decisions require more than two alternatives. For example, you're faced with a menu at a restaurant and want to choose just one of the many delicious options. Applications can encounter the same situation. A user may select only one of the many options from a menu, as an example. The following steps show one method of choosing between multiple options. (The steps assume that you completed the `IfElse` example in the preceding section.) You can also find this function in the `IfElseIf.m` file supplied with the downloadable source code.

1. **In the Editor window, with the `IfElse.m` file selected, click the down arrow under Save and choose Save As from the list that appears.**

   You see the Select File for Save As dialog box. (Refer to Figure 10-2.)

2. **Type IfElseIf.m in the File Name field and click Save.**

   MATLAB saves the example using a new name.

3. **Replace the `IfElse` function name with `IfElseIf`.**

4. **Add the following code after the first `disp()` function call:**

   ```
   elseif Value == 5
       disp('The input value is equal to 5!');
   ```

   At this point, the code provides separate handling for inputs greater than, equal to, and less than 5.

5. **Modify the third `disp()` function statement to read:**

   ```
   disp('The input value is less than 5!');
   ```

   Many people make the mistake of not modifying everything that needs to be modified by an application change. Because you now have a way of handling inputs equal to five, you must change the message so that it makes sense to the user. Failure to modify statements often leads to odd output messages that serve only to confuse users.

6. **Click Save.**

   The function file is saved to disk.

7. **Type IfElseIf(6) and press Enter in the Command window.**

   You see the following output:

   ```
   The input value is greater than 5!
   ```

8. **Type IfElseIf(5) and press Enter in the Command window.**

   You see the following output:

   ```
   The input value is equal to 5!
   ```

9. **Type IfElseIf(4) and press Enter in the Command Window.**

   You see the following output:

   ```
   The input value is less than 5!
   ```

Using the switch statement

You can create any multiple alternative selection code needed using the `if...elseif` statement. However, you have another good way to make selections. A `switch` statement lets you choose one of a number of options using code that is both easier to read and less time-consuming to type. The result is essentially the same, but the method of obtaining the result is different. The following steps demonstrate how to use a `switch` statement. You can also find this function in the `SimpleSwitch.m` file supplied with the downloadable source code.

1. **Click the arrow under the New entry on the Home tab of the MATLAB menu and select Function from the list that appears.**

   You see the Editor window. (Refer to Figure 10-1.)

2. **Delete** `output_args`.

   The example doesn't provide an output argument, but it does require an input argument.

3. **Change the function name from** `Untitled` **to** `SimpleSwitch`.

   The primary function name must match the name of the file.

4. **Change** `input_args` **to** `Value`.

   The function receives a value from the caller to use in the decision-making process.

5. **Type the following code into the function between the comment and the** `end` **keyword.**

```
switch Value
    case 1
        disp('You typed 1.');
    case 2
        disp('You typed 2.');
    case 3
        disp('You typed 3.');
    otherwise
        disp('You typed something greater than 3.');
end
```

This code specifically compares `Value` to the values provided. When `Value` matches a specific value, the application outputs an appropriate message.

At times, the input value doesn't match the values you expect. In such cases, the `otherwise` clause comes into play. It provides the means for doing something even if the input wasn't what you expected. If nothing else, you can use this clause to tell the user to input an appropriate value.

6. **Click Save.**

   You see the Select File for Save As dialog box. (Refer to Figure 10-2.) Notice that the File Name field has the correct filename entered for you.

7. **Click Save.**

   The function file is saved to disk.

8. **Type** SimpleSwitch(1) **and press Enter in the Command window.**

   You see the following output:

   ```
   You typed 1.
   ```

9. **Type** SimpleSwitch(2) **and press Enter in the Command window.**

   You see the following output:

   ```
   You typed 2.
   ```

10. **Type** SimpleSwitch(3) **and press Enter in the Command window.**

    You see the following output:

    ```
    You typed 3.
    ```

11. **Type** SimpleSwitch(4) **and press Enter in the Command window.**

    You see the following output:

    ```
    You typed something greater than 3.
    ```

Understanding the switch difference

A `switch` provides a short method of making specific decisions. You can't make a generalize decision, such as whether a value is greater than some amount. In order to make a match, the value must equal a specific value. The specific nature of a `switch` means that:

- ✔ The code you write is shorter than a comparable `if...elseif` structure.
- ✔ Others are better able to understand your code because it's cleaner and more precise.

✔ The `switch` statement tends to produce focused code that avoids the odd mixing of checks that can occur when using the `if...elseif` structure.

✔ MATLAB is able to optimize the application for better performance because it doesn't have to change a range of values.

✔ Using an `otherwise` clause ensures that people understand that you didn't anticipate a particular value or that the value is less important.

Deciding between if and switch

Whenever possible, use a `switch` statement when you have three or more options to choose from and you can focus attention on specific options for just one variable. Using the `switch` statement has some significant advantages, as described in the preceding section.

The `if` statement provides you with flexibility. You can use it when a range of values is acceptable or when you need to perform multiple checks before allowing a task to complete. For example, when opening a file, you might need to verify that the file actually does exist, that the user has the required rights to access the file, and that the hard drive isn't full. An `if` statement would allow you to check all these conditions in a single piece of code, making the conditions easier to see and understand.

When writing your own applications, you need to keep in mind that there isn't a single solution to any particular problem. The overlap between the `if` and `switch` statements makes it clear that you could use either statement in many situations. Both statements would produce the same output, so there isn't a problem with using them. However, `if` statements are better when flexibility is required, but `switch` statements are better when precision and speed are important. Choose the option that works best for a particular situation, rather than simply choosing an option that works.

Creating Recursive Functions

Many elegant programming techniques exist in the world, but none are quite so elegant as the recursive function. The concept is simple — you create a function that keeps calling itself until a condition is satisfied, and then the function delivers an answer based on the results of all those calls. This process of the function calling itself multiple times is known as *recursion,* and a function that implements it is a *recursive function.*

The most common recursion example is calculating factorial (n!), where *n* is a positive number. (Calculating a factorial means multiplying the number by each number below it in the hierarchy. For example, 4! is equal to 4*3*2*1 or 24.)

Most examples that show how to create a recursive function don't really demonstrate how the process works. The following steps help you create a recursive function that does demonstrate how the process works. Later in the chapter you see a less involved version of the same code that shows how the function would normally appear. You can also find this function in the Factorial1.m file supplied with the downloadable source code.

1. **Click the arrow under the New entry on the Home tab of the MATLAB menu and select Function from the list that appears.**

   You see the Editor window. (Refer to Figure 10-1.)

2. **Change** output_args **to Result.**

   The function returns a result to each preceding cycle of the call.

3. **Change the function name from** Untitled **to** Factorial1.

   The primary function name must match the name of the file.

4. **Change** input_args **to** Value, Level.

   The Value received is always one less than the previous caller received. The Level demonstrates how Value is changing over time.

5. **Type the following code into the function between the comment and the** end **keyword.**

```
if nargin < 2
    Level = 1;
end

if Value > 1
    fprintf('Value = %d Level = %d\n', Value, Level);
    Result = Factorial1(Value - 1, Level + 1) * Value;
    disp(['Result = ', num2str(Result)]);
else
    fprintf('Value = %d Level = %d\n', Value, Level);
    Result = 1;
    disp(['Result = ', num2str(Result)]);
end
```

This example makes use of an optional argument. The first time the function is called, Level won't have a value, so the application automatically assigns it a value of 1.

The code breaks the multiplication task into pieces. For example, when `Value` is 4, the code needs to multiply it by 3 * 2 * 1. The 3 * 2 * 1 part of the picture is defined by the call to `Factorial1(Value - 1, Level + 1)`. During the next pass, `Value` is now 3. To get the appropriate result, the code must multiply this new value by 2 * 1. So, as long as `Value` is greater than 1 (where an actual result is possible), the cycle must continue.

A recursive function must always have an ending point — a condition under which it won't call itself again. In this case, the ending point is the `else` clause. When `Value` is finally less than 1, `Result` is assigned a value of `1` and simply returns, without calling `Factorial1()` again. At this point, the calling cycle unwinds and each level returns, one at a time, until a final answer is reached.

Notice that this example uses a new function, `fprintf()`, to display information onscreen. The `fprintf()` function accepts a formatting specification as its first input. In this case, the specification says to print the string `Value =`, followed by the information found in `Value`, then `Level =`, followed by the information found in `Level`. The `%d` in the format specification tells `fprintf()` to print an integer value. You use `fprintf()` as a replacement for `disp()` when the output formatting starts to become more complex. Notice that `disp()` requires the use of the `num2str()` function to convert the numeric value of `Result` to a string in order to print it.

6. **Click Save.**

   You see the Select File for Save As dialog box. (Refer to Figure 10-2.) Notice that the File Name field has the correct filename entered for you.

7. **Click Save.**

   The function file is saved to disk.

8. **Type** Factorial1(4) **and press Enter in the Command window.**

   You see the following output:

```
Value = 4 Level = 1
Value = 3 Level = 2
Value = 2 Level = 3
Value = 1 Level = 4
Result = 1
Result = 2
Result = 6
Result = 24
ans =
     24
```

The output tells you how the recursion works. Notice that all the `Value` and `Level` outputs come first. The function must keep calling itself until `Value` reaches 1. When `Value` does reach 1, you see the first `Result` output. Of course, `Result` is also 1. Notice how the recursion unwinds. The next `Result` is 2 * 1, then 3 * 2 * 1, and finally 4 * 3 * 2 * 1.

Understanding the fprintf() format specification

A format specification tells a function how to display information onscreen. The `fprintf()` function accepts a regular string as input for a format specification, reading each character. When `fprintf()` encounters a percent (`%`) character, it looks at the next character as a definition of what kind of formatted input to provide. The following list provides an overview of the `%` character combinations used to format information using `fprintf()`.

- ✔ `%bo`: Floating point, double precision (Base 8)

- ✔ `%bu`: Floating point, double precision (Base 10)

- ✔ `%bx`: Floating point, double precision (Base 16, using lowercase letters for the numbers a through f)

- ✔ `%bX`: Floating point, double precision (Base 16, using uppercase letters for the numbers A through F)

- ✔ `%c`: Single character

- ✔ `%d`: Signed integer

- ✔ `%e`: Floating point, exponential notation using a lowercase e

- ✔ `%E`: Floating point, exponential notation using an uppercase E

- ✔ `%f`: Floating point, fixed point notation

- ✔ `%g`: Floating point, general notation using the more compact of %f or %e with no trailing zeros

- ✔ `%G`: Floating point, general notation using the more compact of %f or %E with no trailing zeros

- ✔ `%i`: Signed integer

- ✔ `%o`: Unsigned integer (Base 8)

- ✔ `%s`: String of characters

- ✔ `%to`: Floating point, single precision (Base 8)

- ✔ `%tu`: Floating point, single precision (Base 10)

- ✔ `%tx`: Floating point, single precision (Base 16, using lowercase letters for the numbers a through f)

- ✔ `%tX`: Floating point, single precision (Base 16, using uppercase letters for the numbers A through F)

- ✔ `%u`: Unsigned integer (Base 10)

- ✔ `%x`: Unsigned integer (Base 16, using lowercase letters for the numbers a through f)

- ✔ `%X`: Unsigned integer (Base 16, using uppercase letters for the numbers A through F)

When working with numeric input, you can also specify additional information between the `%` and the subtype, such as `f` for floating point. For example, `%-12.5f`, would display a left-justified number 12 characters in width with 5 characters after the decimal point. See the full details of formatting strings at `http://www.mathworks.com/help/matlab/matlab_prog/formatting-strings.html`.

Now that you have a better idea of how the recursion works, look at the slimmed-down version in Listing 10-1. You can also find this function in the `Factorial2.m` file supplied with the downloadable source code.

Listing 10-1:   A Method for Calculating n!

```
function [ Result ] = Factorial2( Value )
%Factorial2 - Calculates the value of n!
%    Outputs the factorial value of the input number.
    if Value > 1
        Result = Factorial2(Value - 1) * Value;
    else
        Result = 1;
    end

end
```

The final version is much smaller but doesn't output any helpful information to tell you how it works. Of course, this version will run a lot faster, too.

Performing Tasks Repetitively

Giving an application the capability to perform tasks repetitively is an essential part of creating an application of any complexity. Humans don't get bored performing a task once. It's when the task becomes repetitive that true boredom begins to take hold. A computer can perform the same task in precisely the same manner as many times as needed because the computer doesn't get tired. In short, the area in which computers can help humans most is performing tasks repetitively. As with decisions, you have two kinds of structures that you can use to perform tasks repetitively, as described in the sections that follow.

You see a number of terms associated with repetitive code. The same terms that you see used for decision-making code also apply to repetitive code. In addition, the term *loop* is used to describe what repetitive code does. A repetitive structure keeps executing the same series of tasks until such time as the condition for the repetition is satisfied and the loop ends.

Using the for statement

The for statement performs a given task a specific number of times, unless you interrupt it somehow. The examples in the "Making Decisions" section, earlier in this chapter, provide steps for creating functions. Listing 10-2 shows how to use a for loop in an example. You can also find this function in the SimpleFor.m file supplied with the downloadable source code.

Listing 10-2:   Creating Repetition Using the for Statement

```
function [ ] = SimpleFor( Times )
%SimpleFor: Demonstrates the for loop
%    Tell the application how many times to say hello!

    if nargin < 1
        Times = 3;
    end

    for SayIt = 1:Times
        disp('Howdy!')
    end
end
```

In this case, SimpleFor() accepts a number as input. However, if the user doesn't provide a number, then SimpleFor() executes the statement three times by default.

Notice how the variable, SayIt, is created and used. The range of 1:Times tells for to keep displaying the message Howdy! the number of times specified by Times. Every time the loop is completed, the value of SayIt increases by 1, until the value of SayIt is equal to Times. At this point, the loop ends.

Using the while statement

The while statement performs a given task until a condition is satisfied, unless you stop it somehow. The examples in the "Making Decisions" section of the chapter provide steps for creating functions. Listing 10-3 shows how to use a while loop in an example. You can also find this function in the SimpleWhile.m file supplied with the downloadable source code.

Listing 10-3:   Creating Repetition Using the while Statement

```
function [ ] = SimpleWhile( Times )
%SimpleWhile: Demonstrates the while loop
%    Tell the application how many times to say hello!

    if nargin < 1
        Times = 3;
    end

    SayIt = 1;
    while SayIt <= Times
        disp('Howdy!')
        SayIt = SayIt + 1;
    end
end
```

In this example, the function can either accept an input value or execute a default number of times based on whether the user provides an input value for Times. The default is to say "Howdy!" three times.

Notice that the loop code actually begins by initializing SayIt to 1 (so the count begins at the right place). It then compares the current value of SayIt to Times. When SayIt is greater than Times, the loop ends.

You must manually update the counter variable when using a while loop. Notice the line that adds 1 to SayIt after the call to disp(). If this line of code is missing, the application ends up in an *endless loop* — meaning that it never wants to end. If you accidentally create an endless loop, you can stop it by pressing Ctrl+C.

Ending processing using break

It's possible that a loop will ordinarily execute a certain number of times and then stop without incident. However, when certain conditions are met, the loop may have to end early. For example, some people in your organization might be ramping up the Howdy! Application into overdrive. In order to prevent this abuse, you want the loop to stop at five — friendly, but not too verbose. The break clause lets you stop the loop early.

Listing 10-4 shows how to use the break clause with a while loop, but you can use it precisely the same way with the for loop. The examples in the "Making Decisions" section, earlier in the chapter, provide steps for creating functions. You can also find this function in the UsingBreak.m file supplied with the downloadable source code.

Listing 10-4:   Using the break Clause

```
function [ ] = UsingBreak( Times )
%SimpleWhile: Demonstrates the while loop
%    Tell the application how many times to say hello!
%    Don't exceed five times or the application will cut
            you off!

    if nargin < 1
        Times = 3;
    end
```

(continued)

Listing 10-4 *(continued)*

```
    SayIt = 1;
    while SayIt <= Times
        disp('Howdy!')
        SayIt = SayIt + 1;
        if SayIt > 5
            disp('Sorry, too many Howdies')
            break;
        end
    end
end
```

The code executes precisely the same way that the SimpleWhile example works except that this version contains an additional `if` statement. When someone wants to execute the loop more than five times, the `if` statement takes effect. The application displays a message telling the user that the number of Howdies has become excessive and then calls `break` to end the loop. To see this example in action, type **UsingBreak(10)** and press Enter in the Command window.

Ending processing using return

Another way to end a loop is to call `return` instead of `break`. The basic idea is the same. See the upcoming "Differentiating between break and return" sidebar for details on how the two clauses differ.

Listing 10-5 shows how to use the `return` clause with a `while` loop, but you can use it precisely the same way with the `for` loop. The examples in the "Making Decisions" section, earlier in the chapter, provide steps for creating functions. You can also find this function in the `UsingReturn.m` file supplied with the downloadable source code.

Listing 10-5:   Using the return Clause

```
function [ Result ] = UsingBreak( Times )
%SimpleWhile: Demonstrates the while loop
%   Tell the application how many times to say hello!
%   Don't exceed five times or the application will cut
            you off!

    if nargin < 1
        Times = 3;
    end

    Result = 'Success!';
```

```
        SayIt = 1;
        while SayIt <= Times
            disp('Howdy!')
            SayIt = SayIt + 1;
            if SayIt > 5
                disp('Sorry, too many Howdies')
                Result = 'Oops!';
                return;
            end
        end
end
```

Notice that this example returns a Result to the caller. The value of Result is initially set to 'Success!'. However, when the user gets greedy and asks for too many Howdies, the value changes to 'Oops!'. To test this example, begin by typing **disp(UsingReturn())** and pressing Enter. You see the following output:

```
Howdy!
Howdy!
Howdy!
Success!
```

In this case, the application meets with success because the user isn't greedy. Now type **disp(UsingReturn(10))** and press Enter. This time the application complains by providing this output:

```
Howdy!
Howdy!
Howdy!
Howdy!
Howdy!
Sorry, too many Howdies
Oops!
```

Differentiating between break and return

It may seem that break and return are almost precisely the same, but they really aren't. The UsingReturn example demonstrates one difference. You can provide a return value when using the return clause. In some cases, the ability to return a value is essential, and you must use return in place of break.

However, another difference isn't quite so apparent. When you nest one loop inside another, the break clause ends only the inner loop, not the outer loop. As a result, the outer loop continues to run as before. In some cases, this is a desirable behavior, such as when you need to interrupt the current task but still want to execute the remaining tasks. Using return ends all the loops. In fact, the function containing the loops ends and control returns to the caller.

Determining which loop to use

You can understand the `for` and `while` loops better by comparing Listing 10-2 and Listing 10-3. A `for` loop provides the means to execute a set of tasks a precise number of times. You use the `for` loop when you know the number of times a task should execute in advance.

A `while` loop is based on a condition. You use it when you need to execute a series of tasks until the job is finished. (For example, when two functions are expected to converge on a particular value, you can use the `while` loop to detect the convergence and end the processing.) However, you don't know when the task will end until such time as the conditions indicate that the job is done. Because `while` loops require extra code and additional monitoring by MATLAB, they tend to be slower, so you should use the `for` loop whenever possible to create a faster application.

Creating Menus

A menu is one way in which you can start to test the abilities you've gained in this chapter. Listing 10-6 shows a menu that you could use as a model for your own menu. Notice that this menu is a script. You could just as easily create a menu as a function. However, with all the emphasis on functions in this chapter, knowing that you can also use these techniques in scripts is important. You can also find this script in the `MyMenu.m` file supplied with the downloadable source code.

Listing 10-6:   A Simple Script Menu Example

```
EndIt = false;

while not(EndIt)
    clc
    disp('Choose a Fruit');
    disp('1. Orange');
    disp('2. Grape');
    disp('3. Cherry');
    disp('4. I''m Bored, Let''s Quit!');

    Select = input('Choose an option: ');

    if Select == 4
        disp('Sorry to see you go.');
        EndIt = true;
```

```
    else
        switch Select
            case 1
                disp('You chose an orange!');
            case 2
                disp('You chose a grape!');
            case 3
                disp('You chose a cherry!');
            otherwise
                disp('You''re confused, quitting!');
                break;
        end
        pause(2)
    end
end
```

The example begins by declaring a variable, EndIt, to end the while loop. A while loop is the perfect choice in this case because you don't know how long the user will want to use the menu.

The example clears the Command window and then displays the options. After the user enters a selection, the application checks to determine when it should end. If it should, it displays a goodbye message and sets EndIt to true.

When the user chooses some other option, the code relies on a switch to provide a response. In this case, the response is a simple message, but your production application would perform some sort of task. When a user provides a useless response, the application detects it and ends. Notice the use of the break clause.

The pause() function is new. Because the Command window is cleared after each iteration, the pause() function provides a way to display the response for two seconds. The user can also choose to press Enter to return to the menu early.

Part IV

Employing Advanced MATLAB Techniques

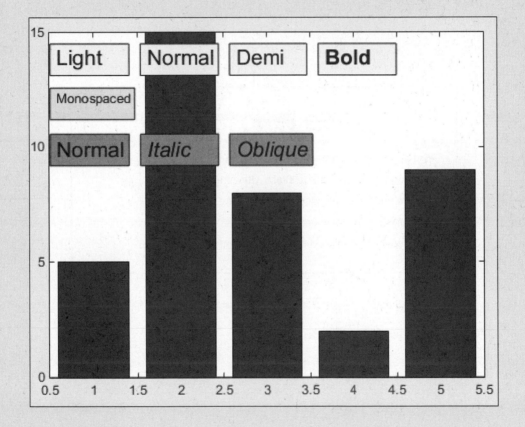

Discover why good comments help you avoid errors at http://www.dummies.com/extras/matlab.

In this part . . .

- ✔ Import data you need to access in MATLAB.
- ✔ Export data you want to share with others.
- ✔ Use fonts and special characters to dress up your plots.
- ✔ Publish and print your MATLAB data.
- ✔ Recover from mistakes you make when using MATLAB.

Chapter 11

Importing and Exporting Data

● ●

In This Chapter
▶ Obtaining data from other sources
▶ Outputting data to other recipients
▶ Interacting with image data

● ●

An application isn't of much use if it can't interact with data — in the case of MATLAB, the variables, formulas, scripts, functions, and plots that you create. In fact, even applications that you might think have nothing to do with data manage quite a lot of it. For example, you might be tempted to think that games don't work with data, but even the lowliest Solitaire game saves statistics, which means that it interacts with data. So you can easily see that most applications interact with at least their own data.

Larger, more complex applications, such as MATLAB, also need some method of interacting with data from other applications. For example, you may need to use Excel data from a buddy to perform a calculation. If MATLAB didn't provide a means to access that data, to *import* it into MATLAB, you couldn't use it to perform the calculation.

After you complete the calculation, you may need to send it back to your buddy, but the only application available at the other site is Excel. Now you must *export* the data from MATLAB into an Excel file that your buddy can use. An Excel data file will help your buddy a lot more than printouts you could send instead because the data is directly accessible.

Plain data — text and numbers — is one thing, but importing and exporting images is quite another. Images are complex because they present graphics — a visual medium — as a series of 0s and 1s. In addition, some image formats have quirks that make them hard to work with. This chapter provides a deeper look into working with image files of various sorts.

Data import and export are among the few activities for which many people find using the GUI easier than typing commands. (The choice you make depends on the complexity of the data and just what you want to achieve by importing or exporting it.) Yes, you can type commands to perform the tasks, and you can add import and export commands to your applications, but importing or exporting complex data manually is often easier using the GUI. This chapter focuses on working with commands. However, you can see how to use the GUI in the "Importing" and "Exporting" sections of Chapter 4. These sections also discuss issues such as which file formats MATLAB supports.

Importing Data

For most people, importing data from various sources is almost a daily chore because our world is based on interconnectivity. Having as much data as possible to perform a task is critical if you want to obtain good results. That's why knowing just how to get the data into MATLAB is so important. It's not just a matter of getting the data, but getting it in such a manner that it can be truly useful. In addition, the import process can't damage the data in any way or it could become useless to you in the long run.

A lot of people get the whole business of the importing and exporting of data confused. *Importing* data always involves taking outside information — something generated externally — and bringing it into a host application, such as MATLAB (as contrasted to *exporting,* which sends information from MATLAB to an external target). So, when your buddy sends you a file with Excel data, you must import it into MATLAB in order to use it.

The following sections describe the essentials for importing data into MATLAB from various sources.

Performing import basics

A basic import uses all the default settings, which works fine for many kinds of data. MATLAB can determine the correct data format relatively often.

An essential part of importing data is to use the correct import function. Each import function has features that make it more suitable to a particular kind of data. Here are some of the text-specific import functions and how they differ:

- ✔ csvread(): Works with numbers only, and the numbers must be separated by commas (hence the name Comma Separated Values, or CSV).

- ✔ dlmread(): Works with numbers only, but the numbers are normally separated by something other than commas.

- ✔ textscan(): Can import both numbers and strings. You must provide a format specification to read the data correctly.

✔ readtable(): Can import both numbers and strings. The output from this function is always a table, even when the source doesn't contain tabular data.

The output you receive depends on the function you use. For example, when working with readtable(), you actually get a table as output, not a matrix or a cell array. On the other hand, using csvread() results in a matrix as output. There are ways to obtain the kind of output you want, but you need to understand that you start with a specific kind of output data from these functions.

The examples found in the sections that follow each use a different method of reading the data from the disk. However, they all use the same data so that you can compare the results. Here's the data found in the NumericData.csv file supplied with the downloadable source code:

```
15,25,30
18,29,33
21,35,41
```

Using csvread()

Using csvread() is the simplest option when working with data of this kind. All you do is type **CSVOutput = csvread('NumericData.csv')** and press Enter. The output is a matrix that contains the following results:

```
CSVOutput =
    15    25    30
    18    29    33
    21    35    41
```

Using dlmread()

The dlmread() function is a little more flexible than csvread() because you can supply a *delimiter* — a character used to separate values — as input. In this case, nothing is odd about the delimiters used in the data file. Each column is separated from the other by a comma. The rows are separated by a newline character. So, all you need to type in this case is **DLMOutput = dlmread('NumericData.csv')**; then press Enter. The output is a matrix containing these results:

```
DLMOutput =
    15    25    30
    18    29    33
    21    35    41
```

Using textscan ( )

The textscan() function can read both strings and numbers in the same data set. However, you must define a format specification to use this function. In addition, you can't simply open the file and work with it. With these requirements in mind, you can use the following steps to help you use the textscan() function.

1. **Type** FileID = fopen('NumericData.csv') **and press Enter.**

   The textscan() function can't open the file for you. However, it does accept the identifier that is returned by the fopen() function. The variable, FileID, contains the identifier used to access the file.

2. **Type** TSOutput = textscan(FileID, '%d,%d,%d/n') **and press Enter.**

   You get a single row of the data as output — not all three rows. So, this is a time when you'd normally use a loop to read the data. However, there is more to see, so the example doesn't use a loop. In this case, the data is read into a cell array, not a matrix.

3. **Type** feof(FileID) **and press Enter.**

   The function outputs a 0, which means that you aren't at the end of the file yet. You might have wondered how you were going to tell the loop to stop reading the file. A simple test using the feof() function takes care of that problem.

4. **Type** TSOutput = [TSOutput; textscan(FileID, '%f,%f,%f/n')] **and press Enter.**

   You now see the second row of data read in. However, look at the format specification. These numbers are read as floating-point values rather than integers. Using textscan() gives you nearly absolute control over the appearance of the data in your application.

5. **Type** isinteger(TSOutput{1,1}) **and press Enter.**

   The output value of 1 tells you that the element at row 1, column 1 is indeed an integer.

6. **Type** isinteger(TSOutput{2,1}) **and press Enter.**

   This step verifies that the element at row 2, column 1 isn't an integer because the output value is 0. It pays to ensure that the data you have in MATLAB is the type you actually expected.

7. **Type** TSOutput = [TSOutput; textscan(FileID, '%2s,%2s,%2s/n')] **and press Enter.**

   This time, the data is read in as individual strings. However, notice that the format specification includes a field width value. If you had simply told textscan() to read strings, it would have read the entire row as a single string into one cell.

8. **Type** textscan(FileID, '%d,%d,%d/n') **and press Enter.**

   This read should take you past the end of the file. The output is going to contain blank cells because nothing is left to read.

9. **Type** feof(FileID) **and press Enter.**

   This time, the output value is 1, which means that you are indeed at the end of the file.

10. **Type** fclose(FileID) **and press Enter.**

    MATLAB closes the file.

    Failure to close a file can cause memory leaks and all sorts of other problems. Not closing the file could quite possibly cause data loss, access problems, or a system crash. The point is that you really don't want to leave a file open after you're done using it.

Now that you have a better idea of how a textscan() should work, it's time to see an application that uses it. Listing 11-1 shows how you might implement the preceding procedure as a function. You can also find this function in the UseTextscan.m file supplied with the downloadable source code.

Listing 11-1:   Using textscan() in an Application

```
function [ ] = UseTextscan( )
%UseTextscan: A demonstration of the textscan() function
%   This example shows how to use textscan() to scan
%   the NumericData.csv file.

    FileID = fopen('NumericData.csv');
    TSOutput = textscan(FileID, '%d,%d,%d/n');

    while not(feof(FileID))
        TempData = textscan(FileID, '%d,%d,%d/n');

        if feof(FileID)
            break;
        end

        TSOutput = [TSOutput; TempData];
    end

    disp(TSOutput);
    fclose(FileID);

end
```

You have already used most of this code as you worked through the exercise, but now you see it all put together. Notice that you must verify that you haven't actually reached the end of the file before adding the data in `TempData` to `TSOutput`. Otherwise, you end up with the blank row that `textscan()` obtains during the last read of the file.

Using readtable ( )

The `readtable()` function works with both strings and numbers. It's a lot easier to use than `textscan()`, but it also has a few quirks, such as assuming that the first row of data is actually column names. To use `readtable()` with the `NumericData.csv` file, type **RTOutput = readtable('NumericData. csv', 'ReadVariableNames', false)** and press Enter. You see the following output:

```
RTOutput =
    Var1      Var2      Var3

    15        25        30
    18        29        33
    21        35        41
```

The output actually is a table rather than a matrix or a cell array. The columns have names attached to them, as shown in Figure 11-1. As a consequence, you can access individual members using the variable name, such as `RTOutput{1, 'Var1'}`, which outputs a value of 15 in this case.

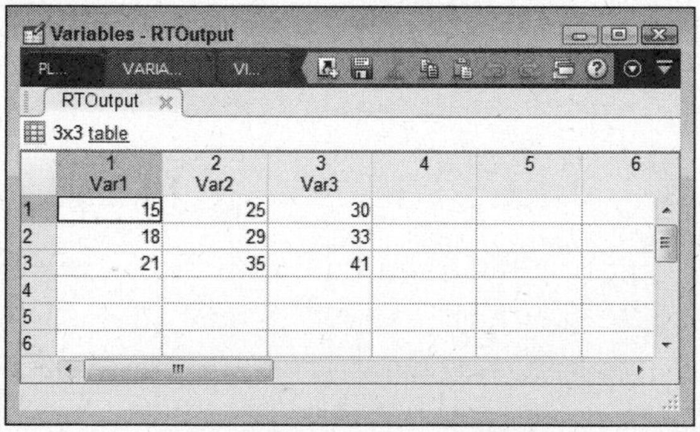

Figure 11-1: Tables provide names for each of the columns.

Notice that `readtable()` accepts property name and value pairs as input. In this case, `'ReadVariableNames'` is a property. Setting this property to `false` means that `readtable()` won't read the first row as a heading of variable names. You use `readtable()` where the output file does contain variable names because having them makes accessing the data easier in many situations.

Importing mixed strings and numbers

Life isn't all about numbers. In some situations, you need to work with a mix of strings and numbers. However, only some of the import functions actually work with strings and numbers. The two that you most commonly use are `textscan()` and `readtable()`. Each function has its own particular capability. For example, `textscan()` provides absolute control over how the data is converted (as described in the "Using `textscan()`" section, earlier in this chapter).

The `readtable()` function is designed more for work with database output, for which the output file likely has header names. The database could reside in a Database Management System (DBMS) or as part of a spreadsheet. The source of the data doesn't matter — only the format does. For this example, you have an output file that contains both row and column headings as shown here (you can also find this data in the `MixedData.csv` file supplied with the downloadable source code):

```
ID,Name,Age,Married
1234,Sam,42,TRUE
2345,Sally,35,TRUE
3456,Angie,22,FALSE
4567,Dan,55,FALSE
```

The first column isn't named because it actually contains the row headers. The `readtable()` function actually has features to handle extras like row headers. Here's a quick overview of the names of properties that you can include in your `readtable()` function call, along with the common values:

- ✔ `FileType`: Defines the type of file. The two acceptable values are `text` and `spreadsheet`.

- ✔ `ReadVariableNames`: Specifies whether the first row contains variable names. The acceptable values are `true` (default), `false`, `1`, or `0`.

- ✔ `ReadRowNames`: Specifies whether the first column contains row names. The acceptable values are `true`, `false` (default), `1`, or `0`.

- ✔ `TreatAsEmpty`: Assigns strings as empty values (such as N/A). You can provide either a single string or a cell array of strings.

- ✔ `Delimiter`: Defines which characters are used as delimiters. You specify this value as a string of individual delimiter characters.

- ✔ `HeaderLines`: Indicates the number of lines to skip at the beginning of the file because they are header lines. The acceptable values are: `0` (default) or any positive integer.

✔ Format: Defines the format of each column using one or more conversion specifiers. The default is to use `double` for numeric data, unless the column contains nonnumeric data, in which case all data appears as a `string`. However, you could change the default numeric value to an `int32` by supplying the `%d` specifier. You can see the details about using the format specifiers at `http://www.mathworks.com/help/matlab/ref/textscan.html#inputarg_formatSpec`.

✔ Sheet: Indicates which worksheet to read in the file. The acceptable values are `1` (default), any positive integer indicating the worksheet index, or a string containing the worksheet name. The worksheets are read one at a time, so you need multiple calls to read multiple worksheets.

✔ Range: Specifies the rectangular portion of the worksheet to read. This value is supplied as a string.

✔ Basic: Determines whether `readtable()` reads the data source in Basic mode (without making any assumptions as to content). The acceptable values are: `true`, `false` (default), `1`, or `0`.

To see `readtable()` in action with the `MixedData.csv` file, type **MixedData = readtable('MixedData.csv', 'ReadRowNames', true, 'Format', '%d%s%d%s')** and press Enter. You see the following output:

```
MixedData =
              Name        Age      Married

     1234    'Sam'        42       'TRUE'
     2345    'Sally'      35       'TRUE'
     3456    'Angie'      22       'FALSE'
     4567    'Dan'        55       'FALSE'
```

Notice that the columns have the appropriate names and that each row has the expected identifier. The only thing missing from the file is `ID`, which is the name of the row identifiers (and is therefore unnecessary).

The function call includes two name and value pairs. The first pair, `'ReadRowNames', true` tells `readtable()` to treat the first column in each row as a row identifier and not as data. The second pair, `'Format', '%d%s%d%s'` provides `readtable()` with the format you want for each column. The formatting options are the same as those used with `textscan()`.

The table has a few interesting features. For example, type **MixedData('1234', 'Age')** and press Enter. You see the following output:

```
ans =
              Age

     1234     42
```

The output is actually a table that contains just the value you want. Notice the use of parentheses for the index. Using identifiable names rather than numeric indexes is also quite nice.

However, you can access the information as actual data rather than as a table. Type **MixedData{'1234', 'Age'}** and press Enter. In this case, you obtain a simple output of 42. The use of curly braces means that you get a data value rather than a table as output. It is still possible to use numeric indexes if you want. Type **MixedData{1, 2}** and press Enter. You obtain the same output value of 42 as you did before. Opening the MixedData variable shows that MATLAB uses both forms of row and column identification, as shown in Figure 11-2.

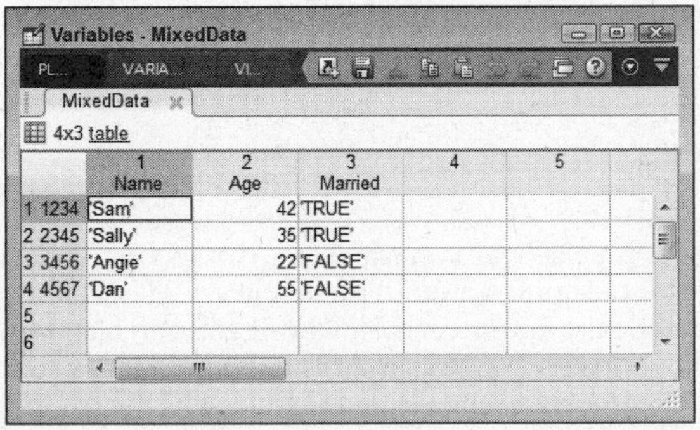

Figure 11-2:
You can use either numeric or text identifiers for tables.

Normally, readtable() treats all numeric data as doubles. The Format property requests an integer value as output instead. Type **class(MixedData{'1234', 'Age'})** and you see an output of int32, so the output is of the correct type.

Defining the delimiter types

Delimiters can cause all kinds of woe. Not every application sticks with the common delimiters, and importing the data can be really hard unless everyone agrees to a single set of rules. The example in this section doesn't follow the rules. Here's the data found in the Delimiters.csv file supplied with the downloadable source code:

```
ID;Name;Age;Married
1234;"Sam";42;TRUE
2345;"Sally";35;TRUE
3456;"Angie";22;FALSE
4567;"Dan";55;FALSE
```

To see just how bad things are, type **MixedData = readtable('Delimiters.csv',
'ReadRowNames', true)** and press Enter. The output you see is

```
MixedData =
   empty 0-by-0 table
```

Essentially, MATLAB is telling you that it doesn't understand the file because
the file contains a mix of data types. You, as a human, can probably look at
the information and figure it out just fine, but MATLAB needs a little more
help. So, add a delimiter specification to the command. Type **MixedData =
readtable('Delimiters.csv', 'ReadRowNames', true, 'Delimiter', ';')** and press
Enter. This time you get the following output:

```
MixedData =
                Name      Age      Married
                ____      ___      _____

        1234    'Sam'     42       'TRUE'
        2345    'Sally'   35       'TRUE'
        3456    'Angie'   22       'FALSE'
        4567    'Dan'     55       'FALSE'
```

This time the output is precisely as you expected. Notice that the output
doesn't include the double quotes around the names. MATLAB removes the
double quotes automatically in this case. However, you can also remove double
quotes by specifying %q instead of %s in a format string. The %q specifier tells
MATLAB to remove the double quotes from the output. If the example file had
used single quotes instead of double quotes around the names , you'd need to
use the textread() function to remove them because the textread() function accepts multiple delimiter characters.

Importing selected rows or columns

Sometimes you don't need an entire file, only certain rows and columns of it.
All four of the functions described earlier in the chapter provide some means
of selecting specific information, but the csvread() function provides a
straightforward example of how to perform this task.

To see just a range of data displayed, type **CSVOutput = csvread('NumericData.
csv', 0, 0, [0, 0, 1, 1])** and press Enter. You see the following output:

```
CSVOutput =
     15     25
     18     29
```

The first argument to csvread() is the name of the file to read. The second
and third arguments are the row and column to start reading. In this case, the
example starts with row 0 and column 0. The fourth argument is a matrix that

specifies the range of values to read. The first two values in the matrix must match the second and third argument because they specify the starting point of the range. The second two values provide the ending point of the range. To give you a better idea of precisely how the range feature works, type **CSVOutput = csvread('NumericData.csv', 0, 1, [0, 1, 2, 2])** and press Enter. This time the output changes to show the second and third columns of the data in `NumericData.csv`:

```
CSVOutput =
    25    30
    29    33
    35    41
```

The row and column values that you use with `csvread()` are zero based. This means that the first row is actually row 0 and the first column is actually column 0. A three-row table would have rows 0 through 2, not 1 through 3. Likewise, a three-column table would contain columns 0 through 2, not 1 through 3.

Exporting Data

After you perform the calculations that you want to perform, you often need to put them in a form that others can use. However, not everyone has a copy of MATLAB on his or her computer, so you need to export the MATLAB data in some other form. Fortunately, getting the data out in a usable form is actually easier than importing it. The following sections describe how to export data, scripts, and functions.

Performing export basics

Importing data often focuses on getting the right results. For example, you might use `textscan()` or `readtable()` on a comma-separated value file, even though a perfectly usable `csvread()` function exists to perform the task. The goal is to get the data from the `.csv` file into MATLAB and preserve both the content and layout of the original information. However, now that you have data inside MATLAB and want to export it, the goal is to ensure that the resulting file is standardized so that the recipient has minimum problems using it. As a result, you'd use `writetable()` only if the recipient really did require a custom format rather than a standard `.csv` file, or if the MATLAB data was such that you had to use something other than `csvwrite()`. Because of the difference in emphasis, the following sections of the chapter focus on standardized export techniques.

Working with matrices and numeric data

Before you can do anything with exporting, you need data to export. Type **ExportMe = [1, 2, 3; 4, 5, 6; 7. 8, 9]** and press Enter. You see the following result:

```
ExportMe =
       1        2        3
       4        5        6
       7        8        9
```

The result is a matrix of three rows and three columns. Exporting matrices is simple because the majority of the functions accept a matrix as a default. To see how exporting matrices works, type **csvwrite('ExportedData1.csv', ExportMe)** and press Enter. MATLAB creates the new file, and you see it appear in the Current Folder window. When you open the file, you see something like the output shown in Figure 11-3. (What you see precisely will vary depending on the application you use to view .csv files.)

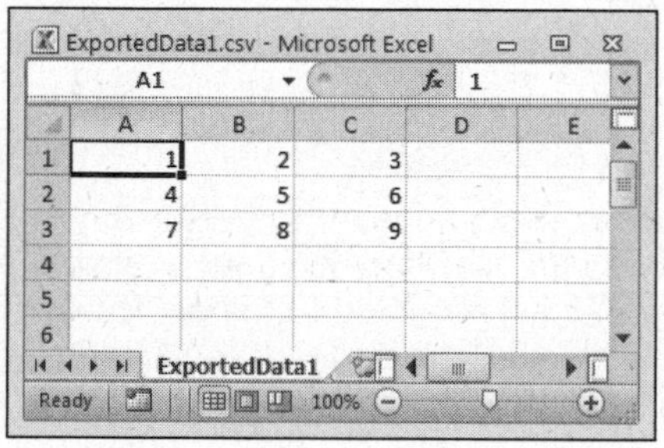

Figure 11-3:
Matrices
provide
the easiest
source for
export.

Not all MATLAB data comes in a convenient matrix. When you use csvwrite(), you must supply a matrix. To get a matrix, you may have to convert the data from the existing format to a matrix using a conversion function. For example, when the data appears as a cell array, you can use the cell2mat() function to convert it. However, some conversions aren't so straightforward. For example, when you have a table as input, you need to perform a two-step process:

1. Use the table2cell() function to turn the table into a cell array.

2. Use the cell2mat() function to turn the cell array into a matrix.

Working with mixed data

Exporting simple numeric data is straightforward because you have a number of functions to choose from that create the correct formats directly. The problem comes when you have a cell array or other data form that doesn't precisely match the expected input for `csvwrite()`. To see how mixed data works, start by typing **MyCellArray = {'Andria', 42, true; 'Michael', 23, false; 'Zarah', 61, false}** and pressing Enter. You see the following result:

```
MyCellArray =
    'Andria'        [42]       [1]
    'Michael'       [23]       [0]
    'Zarah'         [61]       [0]
```

The mixed data type is a problem. If the data were all one type, you could use the `cell2mat()` function to convert the cell array to a matrix like this: `MyMatrix = cell2mat(MyCellArray)`. Unfortunately, if you try that route with the data in `MyCellArray`, you see the following error message:

```
Error using cell2mat (line 46)
All contents of the input cell array must be of the same
        data type.
```

To obtain the required `.csv` file output, you must first convert the cell array into something else. The easiest approach is to rely on a table. Type **MyTable = cell2table(MyCellArray)** and press Enter. You obtain the following output:

```
MyTable =
    MyCellArray1      MyCellArray2      MyCellArray3

    'Andria'          42                true
    'Michael'         23                false
    'Zarah'           61                false
```

At this point, you can type **writetable(MyTable, 'ExportedData2.csv', 'WriteVariableNames', false)** and press Enter. The output will use commas as delimiters between columns, so most applications will see the resulting file as true `.csv` format. Figure 11-4 shows what the output looks like in Excel. (Your output may vary from that shown in the screenshot based on the application you use to view it.)

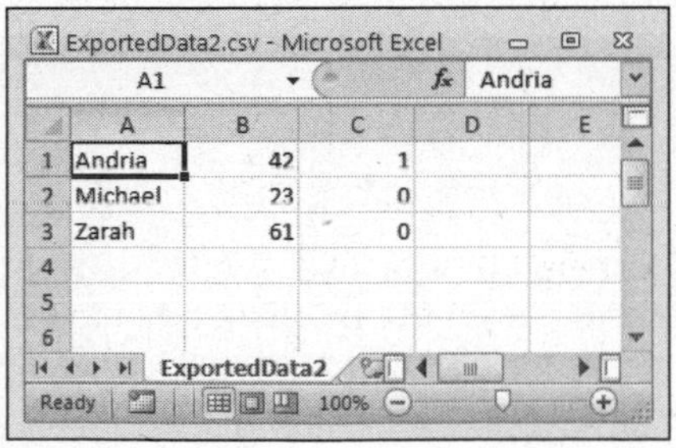

Figure 11-4: Mixed data can be harder to export.

In this case, the exporting process works fine. However, you can always use properties to fine-tune the output of `writetable()`, just as you do with `readtable()`. (See the "Importing mixed strings and numbers" section, earlier in the chapter, for details.) Here is a quick overview of the `writetable()` properties and their uses:

- ✔ `FileType`: Defines the type of file. The two acceptable values are `text` and `spreadsheet`.

- ✔ `WriteVariableNames`: Specifies whether the first row of the output file contains variable names used in MATLAB. The acceptable values are `true` (default), `false`, `1`, or `0`.

- ✔ `WriteRowNames`: Specifies whether the first column of the output file contains the row names used in MATLAB. The acceptable values are `true`, `false` (default), `1`, or `0`.

- ✔ `Delimiter`: Defines which characters are used as delimiters. You specify this value as a string of individual delimiter characters.

- ✔ `Sheet`: Indicates which worksheet to write in the file. The acceptable values are `1` (default), any positive integer indicating the worksheet index, or a string containing the worksheet name. The worksheets are written one at a time, so you need multiple calls to write multiple worksheets.

- ✔ `Range`: Specifies the rectangular portion of worksheet to write. When the MATLAB data exceeds the size of the range, the data is truncated and only the data that will fit appears in the output file. This value is supplied as a string.

Exporting scripts and functions

To export scripts and functions, you must actually publish them using the `publish()` function. MATLAB supports a number of output formats for this purpose. For example, if you want to publish the `UseTextscan()` function that appears earlier in the chapter in HTML format, you type **publish('UseTextscan.m', 'html')** and press Enter. MATLAB provides the following output:

```
ans =
C:\MATLAB\Chapter11\html\UseTextscan.html
```

The actual location varies by system, but you also obtain the location of the published file. Notice that MATLAB places the published file in an `html` subdirectory. Figure 11-5 shows typical output. (What you see may differ based on your platform and the browser that you use.)

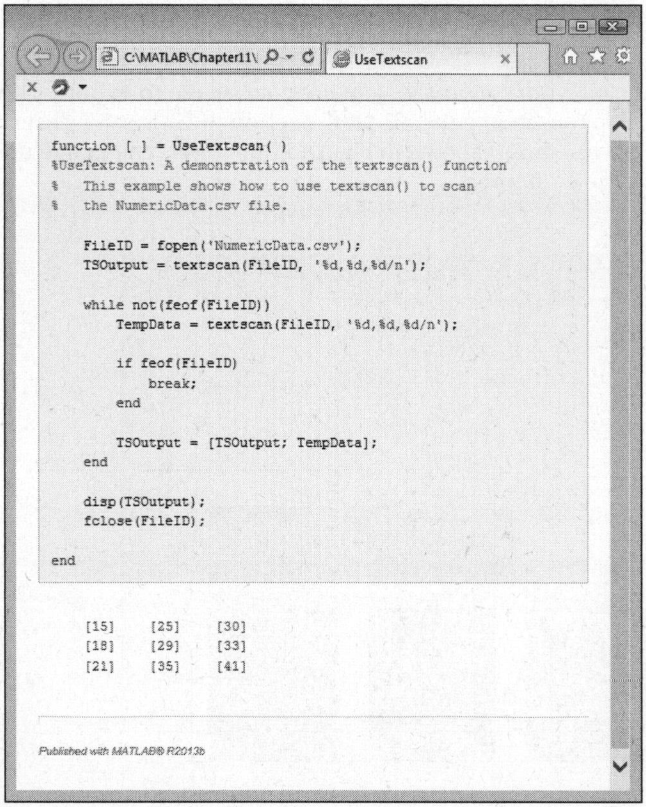

Figure 11-5:
To export
scripts and
functions,
you must
publish
them.

```
function [ ] = UseTextscan( )
%UseTextscan: A demonstration of the textscan() function
%   This example shows how to use textscan() to scan
%   the NumericData.csv file.

    FileID = fopen('NumericData.csv');
    TSOutput = textscan(FileID, '%d,%d,%d/n');

    while not(feof(FileID))
        TempData = textscan(FileID, '%d,%d,%d/n');

        if feof(FileID)
            break;
        end

        TSOutput = [TSOutput; TempData];
    end

    disp(TSOutput);
    fclose(FileID);

end
```

```
    [15]    [25]    [30]
    [18]    [29]    [33]
    [21]    [35]    [41]
```

Published with MATLAB® R2013b

Publishing is a much larger topic than can fit in a single section of a chapter. Chapter 12 discusses publishing in considerably greater detail.

Working with Images

Images are more complex than text files because they use binary data that isn't easy for humans to understand, and the format of that data is intricate. Small, hard-to-diagnose errors can cause the entire image to fail. However, the process of exporting and importing images is relatively straightforward, as described in the following sections.

Exporting images

Before you can export an image, you need an image to export. The "Using the bar( ) function to obtain a flat 3D plot" section of Chapter 7 describes how to create the 3D bar graph shown in Figure 11-6. This is the image used for the remainder of this chapter.

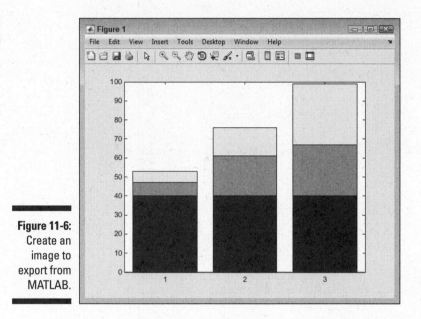

Figure 11-6:
Create an
image to
export from
MATLAB.

Before you export the image, you must decide on the parameters for the output. The most important parameter is the output type. Because Joint Photographic Experts Group (.jpeg) files are common on most platforms, the example uses a .jpeg. However, you can use any of the file formats listed in the Image section of the chart at http://www.mathworks.com/help/matlab/import_export/supported-file-formats.html.

After you decide on an export format, you can use the saveas() function to perform the task. In this case, you type **saveas(gcf( ), 'Bar1.jpeg', 'jpg')** and press Enter. MATLAB exports the figure using whatever resolution is currently set for the figure. Remember that the gcf() function obtains the handle for the current figure. Figure 11-7 shows the plot in Windows Photo Viewer as Bar1.jpeg.

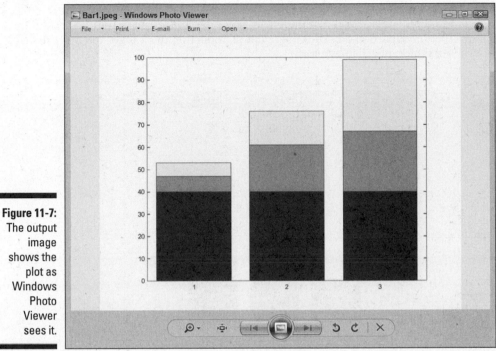

Figure 11-7:
The output
image
shows the
plot as
Windows
Photo
Viewer
sees it.

Use the `saveas()` function to save MATLAB objects, such as plots. However, when working with actual images, use the `imwrite()` function instead. The `imwrite()` function works essentially the same way that `saveas()` does, but it works directly with image files.

Importing images

MATLAB can also work with images that you import from other sources. The basic method of importing an image is to use `imread()`. For example, to import `Bar1.jpeg`, you type **ImportedImage = imread('Bar1.jpeg');** and press Enter. What you see as output is a matrix that has the same dimensions as the image. If the image has a resolution of 900 × 1200, the matrix will also be 900 × 1200. However, a third dimension is involved — the color depth. It takes red, green, and blue values to create a color image. So the resulting matrix is actually 900 × 1200 × 3.

This is one of those situations in which the semicolon is absolutely essential at the end of the command. Otherwise, you may as well go get a cup of coffee as you wait for the numbers to scroll by. If you accidentally issue the command without the semicolon, you can always stop it by pressing Ctrl+C.

To display your image, you use the image() function. For example, to display the image you just imported, you type **image(ImportedImage)** and press Enter. Figure 11-8 shows the result of importing the image. You see the original plot, but in image form.

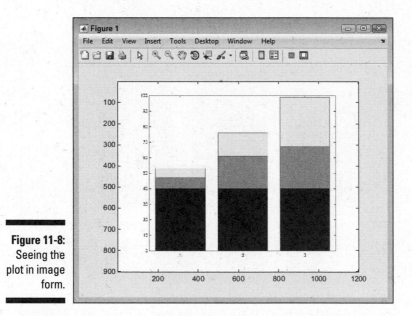

Figure 11-8: Seeing the plot in image form.

Chapter 12

Printing and Publishing Your Work

● ●

In This Chapter

▶ Modifying the appearance of text

▶ Creating a MATLAB data publication

▶ Sending your data to the printer

● ●

After you have labored long to create the next engineering marvel or to produce the next scientific breakthrough, you really do want others to know about it. Most people want to shout about their achievements from the mountaintops, but this chapter tells you only how to publish and print them. *Publishing* is electronic output of your data; *printing* is the physical output of your data. Whether you publish or print your data, it becomes available in a form that other people can use, even if they don't own a copy of MATLAB.

Part of the publishing and printing process is to the make the information aesthetically appealing. It doesn't matter how impressive the data is if no one can read it. Formatting information is part of the documentation process because people need cues about the purpose of the text they're seeing. To that end, this chapter helps you create nicely formatted electronic publications and printed documents.

Using Commands to Format Text

Presentation is a large part of how people receive and understand the material you provide. The same information, presented in two different ways, will elicit different responses from the audience. Little things, such as adding bold type to certain elements, can make a subtle but important difference in how the information is grasped by the audience.

This book isn't about desktop publishing or creating professional-level presentations. The following sections provide you with some ideas on how you can dress up the appearance of your data so that an audience can see the data in a certain light. You can use various text effects to add emphasis where needed or to point out a particular element of interest. In addition, the use of special symbols lets you add explanatory text so that the data is more easily understood.

Modifying font appearance

The fonts you use or misuse can say a lot about your presentation. Unfortunately, fonts are more often misused than used effectively to convey ideas. Here are the four ways in which you can modify the appearance of fonts in MATLAB to good effect:

- ✔ **Bold:** Adds emphasis so that the viewer sees the affected text as being more important than the text around it. However, bold is also used for headings to provide separation between elements. Always think of bold as grabbing the attention of the viewer in some way.

- ✔ **Monospace:** Creates an environment in which text elements line up, as when using a typewriter. Most people use monospace for code or numeric data presented free-form as a table because it helps present a neat appearance. The viewer sees the data rather than being distracted by the data's lack of alignment.

- ✔ **Italic:** Defines elements that are special in some way but don't require emphasis. For example, if you define a term inline with the remainder of your data, the term should appear in italics to cue the viewer that it is explained in the material that follows. The point of italic type is to provide your viewer with cues about added material rather than to add emphasis.

- ✔ **Underline:** Provides pointers to additional resources or other external information (such as the URLs found in Web pages). Some people use an underline for emphasis as well, but this is actually a misuse of an underline because you already have bold to emphasize something. Combining bold and underline is even worse because the recipient can perceive it as shouting. Use underline as a means of pointing to other data that the viewer should know about but that doesn't appear in your presentation.

You can probably find other opinions as to how to use various font styles, but this is the approach used by many (if not most) technical documents, so your viewer will already know the rules. Keeping things familiar and easy to understand will help the viewer focus on your data rather than on the fonts and styles you use. Always make the data king of the presentation and leave the beautiful text to the artists of the world.

Listing the available fonts

At some point, you need to know how to obtain a list of fonts available on the local system. To do so, you use the listfonts() function. To see how this function works, type **Fonts = listfonts();** and press Enter. The Fonts variable receives a sorted list of fonts on the current system that you can then search for appropriate values. To find a specific font, such as Arial, type **Found = find(strcmp(Fonts, 'Arial'));** and press Enter. If Found isn't empty after the call (in other words, length(Found) returns a value greater than 0), the system supports the font you want to use.

Some elements can have special fonts in MATLAB. To discover the identity of the special font, you call listfonts() with the handle of the element. For example, a user interface element may have a special font. If you obtain a handle to that user interface element and pass it to the listfonts() function, you receive not only a sorted list of the system fonts but also the identity of the special font used in the user interface element.

Now that you have some idea of how to use the font styles, the following sections demonstrate how to add them to your presentation. The screenshots in each section build on the section before it so that you can see all the effects in action. You don't have to work through the sections in any particular order.

Bold

The use of emphasis, normally associated with bold type, can make data stand out. However, in MATLAB, the term *bold* actually refers to font weight. The strength of the font you use provides a level of emphasis. In fact, you can set a font to four different levels of emphasis:

- ✔ Light
- ✔ Normal (default)
- ✔ Demi
- ✔ Bold

Not every font that you have installed on your system supports all four levels of emphasis, but at least some do. You may try to achieve a certain level of emphasis with a font and find that your efforts are thwarted. In many cases, it has nothing to do with your code and everything to do with the font you're using. With these font limitations in mind, the following steps help you see the varying levels of emphasis that you can achieve using MATLAB.

1. **Type** Bar1 = bar([5, 15, 8, 2, 9]); **and press Enter.**

   MATLAB creates a new bar chart, as shown in Figure 12-1.

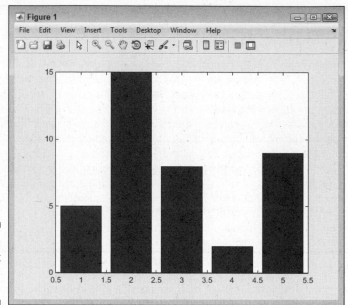

Figure 12-1:
A bar chart
that you can
format.

2. **Type** TBox1 = annotation('textbox', [.13, .825, .14, .075], 'String', 'Light', 'FontName', 'Arial', 'FontSize', 16, 'FontWeight', 'light', 'BackgroundColor', [1, 1, 0]); **and press Enter.**

   You see a new annotation of type `textbox` added to the plot. The various entries that you typed change the default font, the font size (so that you can more easily see the text), the font weight (as a matter of emphasis), and the background color (so that the textbox stands out from the bars in the background). To see the various weights side by side, the next few steps add three more textboxes, each with a different font weight.

3. **Type** TBox2 = annotation('textbox', [.29, .825, .14, .075], 'String', 'Normal', 'FontName', 'Arial', 'FontSize', 16, 'FontWeight', 'normal', 'BackgroundColor', [1, 1, 0]); **and press Enter.**

   You see the next annotation placed on the second bar. In most cases, you won't see much (or any) difference between the light and normal settings because few fonts support both light and normal. However, some do, so it's important to experiment.

4. **Type** TBox3 = annotation('textbox', [.45, .825, .14, .075], 'String', 'Demi', 'FontName', 'Arial', 'FontSize', 16, 'FontWeight', 'demi', 'BackgroundColor', [1, 1, 0]); **and press Enter.**

5. **Type** TBox4 = annotation('textbox', [.61, .825, .14, .075], 'String', 'Bold', 'FontName', 'Arial', 'FontSize', 16, 'FontWeight', 'bold', 'BackgroundColor', [1, 1, 0]); **and press Enter.**

You see the final annotation placed on the plot. Looking at the four different annotations, you can see a progression of weights and emphasis. Even though you may not see a difference between light and normal, the weight differences among normal, demi, and bold are pronounced, as shown in Figure 12-2.

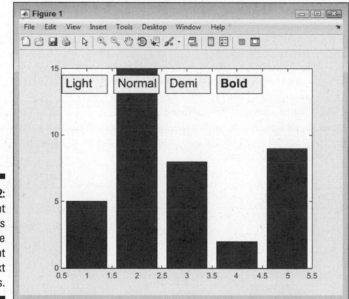

Figure 12-2:
Different font weights provide different levels of text emphasis.

Monospace

A monospace font is one in which the characters take up precisely the same amount of space. It's reminiscent of the kind of text that typewriters put out (if you don't know what a typewriter is, check out `http://site.xavier.edu/polt/typewriters/tw-history.html`). The reason that monospace is still useful is because getting elements to line up is easy. That's why code is often presented in monospace: You can easily see the indentation that the code needs to look nice and work properly. The way that monospace is set for an element is through the font. Here are some commonly used monospace fonts:

✔ Anonymous Pro

✔ Courier

- ✔ Courier New
- ✔ Fixedsys
- ✔ Letter Gothic
- ✔ Lucida Sans Typewriter Regular
- ✔ Lucida Console
- ✔ Monaco
- ✔ Profont
- ✔ Ubuntu

You can see some additional monospace fonts at http://www.fontsquirrel.com/fonts/list/classification/monospaced. The point is to use a font that provides the proper appearance for your application. To obtain a mono-space font appearance, you simply change the FontName property to a mono-space font. For example, type **TBox5 = annotation('textbox', [.13, .72, .15, .075], 'String', 'Monospaced', 'FontName', 'CourierNew', 'BackgroundColor', [0, 1, 1]);** and press Enter to produce a text box containing monospace text (using the plot generated in the preceding section of the chapter). Figure 12-3 shows the results of this command.

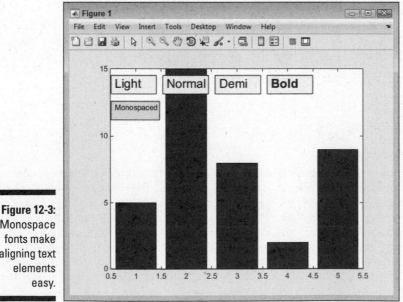

Figure 12-3: Monospace fonts make aligning text elements easy.

Italic

Fonts normally have a straight, up-and-down appearance. However, you can skew the font to give an angle to the upright characters and change how the font looks. The skewed version of a font is called *italic*. To create an italic font, the person creating the font must design a separate set of letters and place them in a file containing the italic version.

Some fonts don't have an italic version. When you encounter such a situation, you can ask the computer to skew the font programmatically. The font file hasn't changed and the font is still straight up and down, but it looks italicized onscreen. This version of italics is called *oblique*. The italic version of a font always gives a better visual appearance than the oblique version because the italic version is hand tuned — that is, individual pixels are modified so that the font appears smoother.

To configure a font for either italic or oblique, you use the FontAngle property. The following steps help you see the differences between the standard, italic, and oblique versions of a font. These steps assume that you have created the MATLAB plot found in the "Bold" section, earlier in this chapter.

1. **Type** TBox6 = annotation('textbox', [.13, .61, .14, .075], 'String', 'Normal', 'FontSize', 16, 'FontAngle', 'normal', 'BackgroundColor', [1, 0, 1]); **and press Enter.**

   You see a textbox annotation containing a normal version of the default MATLAB font for your system.

2. **Type** TBox7 = annotation('textbox', [.29, .61, .14, .075], 'String', 'Italic', 'FontSize', 16, 'FontAngle', 'italic', 'BackgroundColor', [1, 0, 1]); **and press Enter.**

   You see a textbox annotation containing either a normal or an italic version of the default MATLAB font for your system. The italic version appears only if the font happens to have an italic version (most do).

3. **Type** TBox8 = annotation('textbox', [.45, .61, .15, .075], 'String', 'Oblique', 'FontSize', 16, 'FontAngle', 'oblique', 'BackgroundColor', [1, 0, 1]); **and press Enter.**

   You see a textbox annotation containing an oblique version of the standard font. Figure 12-4 shows all three versions. You may see a slight difference in angle between the italic and oblique versions. The oblique version may seem slightly less refined. Then again, you might not see any difference at all between italic and oblique.

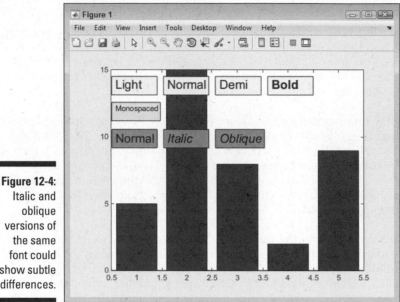

Figure 12-4:
Italic and
oblique
versions of
the same
font could
show subtle
differences.

Underline

Interestingly enough, MATLAB doesn't provide a simple method to underline text. For example, you can't easily perform this particular task using the GUI. In fact, the default method of creating text using code doesn't provide a method for underlining text, either. However, you have a way to accomplish the goal using a special interpreter called LaTeX (http://www.latex-project.org/).

The LaTeX interpreter is built into MATLAB, but it isn't selected by default, so you must set it using the Interpreter property. In addition to underlining text, you use the \underline() LaTeX function. To see how this works, type **TBox9 = annotation('textbox', [.13, .5, .175, .075], 'String', '\underline{Underline}', 'FontSize', 16, 'BackgroundColor', [.5, 1, .5], 'Interpreter', 'latex');** and press Enter. You see the output shown in Figure 12-5.

Notice that the output differs quite a bit when using the LaTeX interpreter. That's because LaTeX ignores many of the formatting properties supplied by MATLAB — you must set them using LaTeX functions. However, the problem is more serious than simply setting a font because MATLAB appears to lack the required font files for LaTeX. The bottom line is that you should avoid underlining text unless you truly need to do so. Use bold, italic, and font colors in place of the underline as often as possible. The discussion at http://www.mathworks.com/matlabcentral/newsreader/view_thread/114116 can provide you with additional information about this issue and some potential work-arounds.

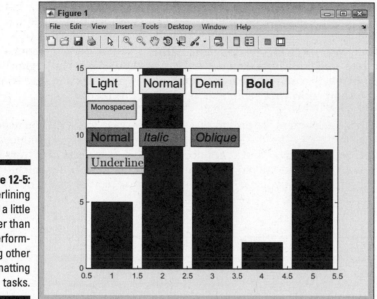

Figure 12-5:
Underlining text is a little harder than performing other formatting tasks.

Using special characters

Sometimes you need to use special characters and character formatting in MATLAB. The following sections describe how to add Greek letters to your output, as well as work with superscript and subscript as needed.

Greek letters

The 24 Greek letters are used extensively in math. To add these letters to MATLAB, you must use a special escape sequence, similar to the escape sequences you use in previous chapters. Table 12-1 shows the sequences to use for Greek letters.

Table 12-1		Adding Greek Letters to MATLAB			
Letter	*Sequence*	*Letter*	*Sequence*	*Letter*	*Sequence*
α	\alpha	β	\beta	γ	\gamma
δ	\delta	ε	\epsilon	ζ	\zeta
η	\eta	θ	\theta	ι	\iota
κ	\kappa	λ	\lambda	μ	\mu
ν	\nu	ξ	\xi	o	Not Used
π	\pi	ρ	\rho	σ	\sigma
τ	\tau	υ	\upsilon	φ	\phi
χ	\chi	ψ	\psi	ω	\omega

As you can see, each letter is preceded by a backslash, followed by the letter's name. The output is always lowercase Greek letters. Notice that omicron (o) has no sequence. To see how the letters appear onscreen, type **TBox10 = annotation('textbox', [.13, .39, .17, .085], 'String', '\alpha\beta\gamma\delta\ epsilon\zeta\eta\theta\iota\kappa\lambda\mu \nu\xi\pi\rho\sigma\tau\ upsilon\phi\chi\psi\omega', 'BackgroundColor', [.5, .5, 1])**; and press Enter. Figure 12-6 shows how your sample should look at this point.

Many of the Greek letters are also available in uppercase form. All you need to do is use initial caps for the letter name. For example, \gamma produces the lowercase letter, but \Gamma produces the uppercase version of the same letter. You can obtain additional information about text properties (including additional symbols that you can use) at http://www.mathworks.com/ help/matlab/ref/text_props.html.

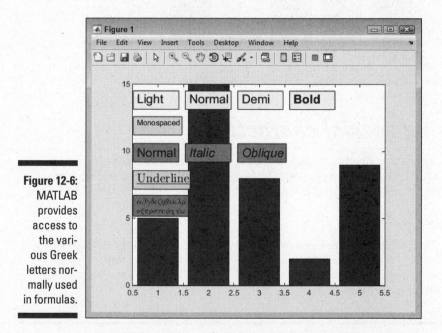

Figure 12-6: MATLAB provides access to the various Greek letters normally used in formulas.

Superscript and subscript

Using superscript and subscript as part of the output is essential when creating formulas or presenting certain other kinds of information. MATLAB uses the caret (^) to denote superscript and the underscore (_) to denote subscript. You enclose the characters that you want to superscript or subscript in curly brackets { }. To see how superscript and subscript works, type **TBox11 = annotation('textbox', [.45, .39, .15, .075], 'String', 'Normal^{Super}_{Sub}', 'BackgroundColor', [.5, .5, 1])**; and press Enter. Figure 12-7 shows typical output from this command.

Notice that the superscript and subscript characters appear in the command without a space after the characters that are in normal type. The output shows these characters immediately after the normal type. In addition, the superscripted characters are over the top of the subscripted characters.

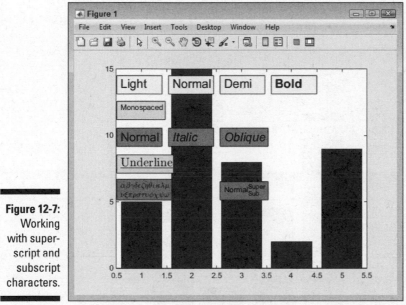

Figure 12-7:
Working with super-script and subscript characters.

Adding math symbols

You would have a tough time presenting formulas to others without being able to use math symbols. MATLAB provides you with a wealth of symbols that you can use for output purposes. The following sections describe the most commonly used symbols and how you access them.

Fraction

Displaying a fraction onscreen doesn't always include a numeric fraction; it could be a formula that requires that sort of presentation. Whatever your need, you can display fractions whenever needed. However, to do that, you must use the LaTeX interpreter mentioned in the "Underline" section, earlier in the chapter. This means that your formatting options are limited and that the output won't necessarily reflect the formatting choices you normally make when using MATLAB.

The fraction requires use of a LaTeX display style that you access using $\displaystyle\frac. The fraction itself appears in two curly brackets, such as {1}{2} for the symbol ½. The entry ends with another dollar sign ($). To see how fractions work with something a little more complex, type **TBox12 = annotation('textbox', [.13, .3, .14, .075], 'String', '$\displaystyle\frac{x-2y}{x+y}$', 'BackgroundColor', [1, .5, .5], 'Interpreter', 'latex');** and press Enter. Figure 12-8 shows typical output from this command.

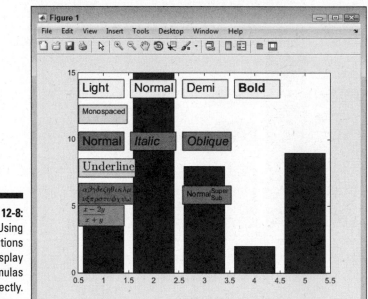

Figure 12-8:
Using fractions to display formulas correctly.

Square root

MATLAB makes displaying a square root symbol easy. However, getting the square root symbol the right size and with the bar extended over the expression whose root is being taken requires LaTeX. As with many LaTeX commands, you enclose the string that you want to format in a pair of dollar signs ($). The function used to perform the formatting is \sqrt{ }, and the value that you want to place within the square root symbol appears within the curly brackets.

To see the square root symbol in action, type **TBox13 = annotation('textbox', [.29, .3, .14, .075], 'String', '$\sqrt{f}$', 'BackgroundColor', [1, .5, .5], 'Interpreter', 'latex');** and press Enter. The variable f will appear in the square root symbol, as shown in Figure 12-9.

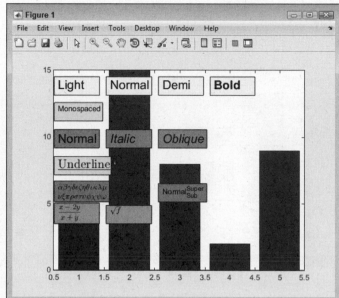

Figure 12-9:
Displaying
the square
root symbol
so that the
bar extends
over the f.

Sum

Displaying a summation formula complete with sigma and the upper and lower limit involves using LaTeX with the \sum function. You supply all three elements of the display in a single statement: the lower limit first, the upper limit second, and the expression third. Each element appears in separate curly brackets. The lower limit is preceded by the underscore used for subscripts and the upper limit is preceded by the caret used for superscripts. The entire statement appears within dollar signs ($), as is normal for LaTeX. However, in this particular case, you must include a second set of dollar signs or the expression doesn't appear correctly onscreen. (The upper and lower limits don't appear in the correct places.)

To see how summation works, type **TBox14 = annotation('textbox', [.45, .285, .14, .1], 'String', '$$\sum_{i=1}^{2n}{|k_i-k_j|}$$', 'BackgroundColor', [1, .5, .5], 'Interpreter', 'latex');** and press Enter. Notice the use of the double dollar signs in this case. In addition, be sure to include both the underscore and caret, as shown. Figure 12-10 shows the result of using this command.

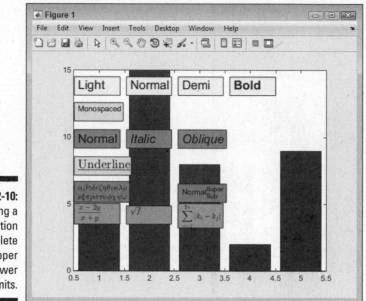

Figure 12-10:
Showing a summation complete with upper and lower limits.

Integral

To display a definite integral, you use the LaTeX \int function, along with the \d function for the slices. The \int function accepts three inputs: two for the interval and the third for the function. In many respects, the format is the same as that used for summation. The beginning of the interval relies on the superscript caret character, while the ending of the interval relies on the subscript underscore character. You must enclose the entire command within double dollar signs ($$) or else the formatting of the superscript and subscript will fail.

To see how to create an integral, type **TBox15 = annotation('textbox', [.61, .285, .22, .1], 'String', '$$\int_{y1(x)}^{y2(x)}{f(x,y)}\d{dx}\d{dy}$$', 'BackgroundColor', [1, .5, .5], 'Interpreter', 'latex');** and press Enter. Notice that the two slices come after the \int function and that each slice appears in its own \d function. Figure 12-11 shows the result of this command.

Derivative

When creating a derivative, you use LaTeX to define a combination of a fraction with superscripts. So, in reality, you've already created a derivative in the past — at least in parts. To see how a derivative works, type **TBox16 = annotation('textbox', [.13, .21, .14, .085], 'String', '$\displaystyle\frac{d^2u} {dx^2}$', 'BackgroundColor', [1, .5, .5], 'Interpreter', 'latex');** and press Enter. Figure 12-12 shows the output from this command.

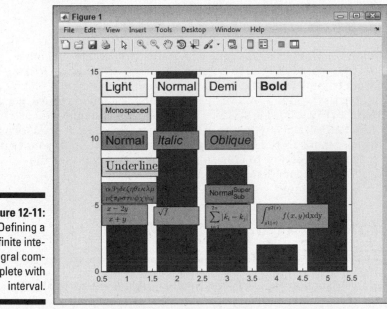

Figure 12-11:
Defining a definite integral complete with interval.

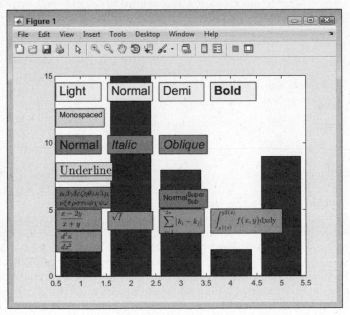

Figure 12-12:
Use a combination of fractions and superscripts to create a derivative.

Publishing Your MATLAB Data

At some point, you want to publish the information you create. Of course, most of the time, you don't need to publish a matrix or other source data. What you want to publish are the plots you create from the data. A picture is worth a thousand words — you've heard the phrase a million times, yet it still holds true. The following sections describe the techniques you use to output MATLAB data of any sort to other formats. However, it does focus on plots because that's the kind of data you output most often. (If you want to discover the basics of how to publish functions and scripts, see the "Exporting scripts and functions" section of Chapter 11.)

Performing advanced script and function publishing tasks

The publish() function can perform a number of different sorts of publishing tasks. The simplest way to publish a script or function is to call publish() with the name of the file. This approach produces an HTML file output. If you want to specify a particular kind of output, you provide the file format as the second input, such as publish('Bar1.m', 'pdf'). MATLAB supports the following output formats:

- ✔ .doc
- ✔ .html (default)
- ✔ .latex
- ✔ .pdf
- ✔ .ppt
- ✔ .xml

After you finish publishing a script or function in HTML format, you can view it using the web() function. To test this feature, first publish the Bar1.m script by typing **publish('Bar1.m')** and pressing Enter. After you see the success message, type **web('html\Bar1.html')** and press Enter. You see the MATLAB browser output, shown in Figure 12-13. However, even if this output looks perfect, always test the published output using the applications that your viewers will use to ensure that everything displays correctly.

Saving your plot as a script

MATLAB provides two command-based methods for saving your plot as a script. Both involve using the saveas() function. The first method creates a complete script for your plot, which makes the code used to create the plot easy to edit. Assuming that you want to save the Bar1 plot used throughout this chapter, you type **saveas(Bar1, 'CreateBar1', 'mmat')** and press Enter. The output is a script containing all the required drawing commands and an associated resource file with a .mat extension. This first method will eventually become obsolete (and you see a message telling you so), but it works fine in the current version of MATLAB.

The second method creates a script for programmatically loading the plot from disk. To perform this task, type **saveas(Bar1, 'Bar1', 'm')** and press Enter. In this case, you get a script for loading the plot from disk. In addition, MATLAB saves a copy of the plot for you. However, it saves just the plot because you used the Bar1 handle. If you want to save the entire figure instead, you must type **saveas(gcf(), 'Figure 1', 'm')** and press Enter (assuming that Figure 1 is indeed the current figure).

Another possibility is to use the GUI to create a function for your plot. To perform this task, choose File⇨Generate Code in the figure window. You see a message telling you that MATLAB is generating the code for you. When the message box disappears, the code is complete, and you see an editor window containing the resulting code. This technique creates a function to regenerate the entire figure. You can also click Edit Plot on the figure toolbar and then right-click individual elements. Choose Show Code from the context menu and you see an editor window containing a function to generate just that element.

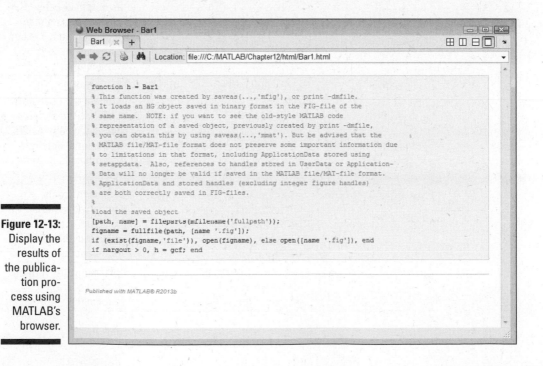

Figure 12-13: Display the results of the publication process using MATLAB's browser.

```
function h = Bar1
% This function was created by saveas(...,'mfig'), or print -dmfile.
% It loads an HG object saved in binary format in the FIG-file of the
% same name.  NOTE: if you want to see the old-style MATLAB code
% representation of a saved object, previously created by print -dmfile,
% you can obtain this by using saveas(...,'mmat'). But be advised that the
% MATLAB file/MAT-file format does not preserve some important information due
% to limitations in that format, including ApplicationData stored using
% setappdata.  Also, references to handles stored in UserData or Application-
% Data will no longer be valid if saved in the MATLAB file/MAT-file format.
% ApplicationData and stored handles (excluding integer figure handles)
% are both correctly saved in FIG-files.
%
%load the saved object
[path, name] = fileparts(mfilename('fullpath'));
figname = fullfile(path, [name '.fig']);
if (exist(figname,'file')), open(figname), else open([name '.fig']), end
if nargout > 0, h = gcf; end
```

Published with MATLAB® R2013b

You can also use a version of publish() with name and value pair options. This version of publish() gives you the most control over the published output. You can control precisely where the document is saved and the size of any images provided with the output. Table 12-2 contains a list of the options and describes how you can use them.

Table 12-2	Using the publish() Options		
Name	*Values*	*Type*	*Description*
catchError	true (default) or false	Code	Determines how MATLAB handles errors during the publishing process.
codeToEvaluate	String containing the required code	Code	Allows you to provide additional code with the published document so that it's possible to perform tasks such as evaluating the code when the associated function requires inputs.
createThumbnail	true (default) or false	Figure	Determines whether the output document contains a thumbnail version of the full image (when an image is part of the output).
evalCode	true (default) or false	Code	Forces MATLAB to evaluate the code as it publishes the script or function, which results in additional details in the output file in some cases.
figureSnapMethod	entireGUIWindow (default), print, getframe, or entireFigure Window	Figure	Defines the technique used to obtain the figure contained within the published document.
format	doc, html (default), latex, pdf, ppt, and xml	Output	Determines the format of the published document.

Name	Values	Type	Description
imageFormat	png, epsc2, jpg, bmp, tiff, eps, eps2, ill, meta, or pdf	Figure	Indicates the format of the figure contained within the published document. The default setting depends on the output document format. Some document formats won't access all the format types. For example, PDF output is limited to the bmp or jpg options, but XML output can accept any of the file formats.
maxHeight	' ' (default) or positive integer value	Figure	Determines the maximum height of the figure contained within the published document.
maxOutputLines	Inf (default) or non-negative integer value	Code	Specifies the number of lines of code that MATLAB includes in the published document. Setting this value to 0 means that no code is in the output.
maxWidth	' ' (default) or positive integer value	Figure	Determines the maximum width of the figure contained within the published document.
outputDir	' ' (default) or full path to output directory	Output	Specifies where to place the published document on disk.
showCode	true (default) or false	Code	Determines whether the published document contains any source code.
stylesheet	' ' (default) or full path and XSL filename	Output	Specifies the location and name of an XSL file to use when generating XML file output.
useNewFigure	true (default) or false	Figure	Specifies that MATLAB is to create a new figure prior to publishing the document.

Saving your figures to disk

You must save your figures to disk if you want to use them the next session. However, saving a figure to disk can also help you publish the information in a form that lets others use the information as well. The format you choose determines how the saved information is used. Only the MATLAB figure (.fig) format provides an editable form that you can work with during the next session. The following sections describe the GUI and command method of saving figures to disk.

Using the GUI to save figures

To save an entire figure, choose File⇨Save As in the figure window. You see the Save As dialog box, shown in Figure 12-14. Type a name for the file in the File Name field, select the format that you want to use to save the file in the Save As Type field, and click Save to complete the process.

Using commands to save figures

The command version of saving a figure depends on the saveas() command. To use this command, you supply a handle to the figure that you want to save as the first argument. The second argument is a filename. When you provide a type of file format to use as the third argument, the filename need not include a file extension. However, if you provide just the filename, you must provide an extension as well. MATLAB supports these file formats:

- .ai
- .bmp
- .emf
- .eps
- .fig
- .jpg
- .m
- .pbm
- .pcx
- .pdf
- .pgm
- .png
- .ppm
- .tif

The handle that you provide need not be the figure itself. For example, if you type **saveas(TBox1, 'TBox1.jpg')** and press Enter, MATLAB still saves the entire figure. (Note that you have no way to specify that you want to save just a portion of the figure.)

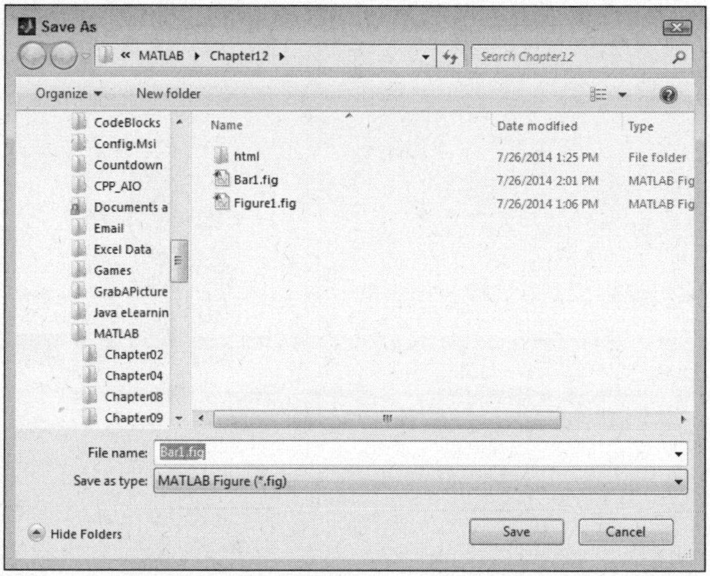

Figure 12-14:
Saving an entire figure.

Printing Your Work

Printing is one of the tasks that most people use the GUI to perform, even the most ardent keyboard user. The issue is one of convenience. Yes, you can use the `printopt()` and `print()` functions to perform the task using the keyboard, but only if you're willing to perform the task nearly blind. The GUI actually shows you what the output will look like (or, at least, a close approximation). Using the commands is significantly harder in this case and isn't discussed in the book. The following sections describe how to use the GUI to output your document.

Configuring the output page

Before you print your document, you need to tell MATLAB how to print it. Choose File➪Print Preview in the figure window to display the Print Preview dialog box, shown in Figure 12-15.

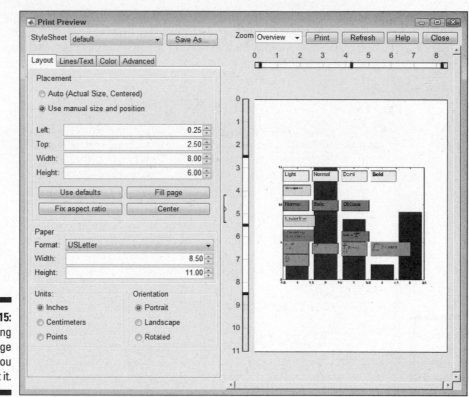

Figure 12-15:
Configuring
the page
before you
print it.

The right side of the dialog box shows an approximation of the changes you make on the left side. The settings are presented on four tabs with the following functions:

✔ **Layout:** Defines how the plot or other document will appear on the page. You specify the margins, size of the paper, and presentation of the information itself. The best way to work through the layout process is simply to try the settings and see how they affect the image shown in the right side of the dialog box.

✔ **Lines/Text:** Specifies the line widths and types of text used to print the documentation. In some cases, you find that using a different line weight or font for printing improves the output appearance of the document. Some of these settings are printer specific. For example, a built-in font is likely to provide a more pleasing appearance than a software font provided as part of the system.

- ✔ **Color:** Fixes the color changes between the display and the printer. What you see onscreen is unlikely to be what you get from the printer. The reasons for the difference are complex, but from an overview perspective, printers use subtractive color mixing, while displays use additive color mixing. The two kinds of color presentation don't always work in sync with each other. Read the article at `http://www.worqx.com/color/color_systems.htm` for additional information.

- ✔ **Advanced:** Provides access to features such as recomputing the links and ticks before printing. You won't generally need these settings unless you notice a problem during the printing process.

Printing the data

After you create an output configuration, you can print the document. To perform this task, choose File➪Print. You see the Print dialog box, in which you can choose a printer, configure it if necessary by clicking the Properties button, and then click OK to print the document.

You don't have to have direct access to the printer you want to use. The Print to File option in the Print dialog box lets you output the printed material to a file on disk. You can then send this file to the printer when you do have access to it. As an alternative, you can also send the file to someone else who needs the printed output (assuming that the person has access to the necessary printer).

Chapter 13

Recovering from Mistakes

· ·

In This Chapter

▶ Understanding error messages

▶ Considering the uses of quick alerts

▶ Obtaining common fixes for MATLAB error messages

▶ Developing your own error messages

▶ Embracing good coding practices

· ·

A lot of people associate mistakes with failure. Actually, everyone makes mistakes, and they're all learning opportunities. Some of the greatest minds the world has ever seen thrived quite nicely on mistakes. (Consider how many tries it took Edison to get the light bulb right; also, see my blog post at http://blog.johnmuellerbooks.com/2013/04/26/defining-the-benefits-of-failure/ on the benefits of failure.) The point is that mistakes are normal, common, and even advantageous at times, as long as you get over them and recover. That's what this chapter is all about — helping you recover from mistakes.

MATLAB can't talk to you directly, so it communicates through error messages. Part of the problem is that MATLAB and humans speak two different languages, so sometimes the error messages aren't quite as helpful as they could be. This chapter helps you understand the error messages so that you can do something about them. You can even find sources for common fixes to these error messages, so the work of overcoming any issues may have already been done by someone. All you'll have to do is go to the right source.

When you write your own applications, sometimes you have to tell other people that they have made a mistake. Unfortunately, you can't record your voice saying, "Oops, I think you meant to type a number instead of some text here" and expect MATLAB to play it for someone using your application. Therefore, you need to create your own custom error messages so that you can communicate with others using your applications.

Of course, the best way to handle error messages is to not make mistakes. Even though some mistakes are unavoidable (and others have had positive benefits, such as the making of the glue found on the back of sticky notes), you can work in ways that tend to reduce errors of all sorts. The final section of this chapter discusses good coding practices to use to avoid trouble. No, this section can't prevent you from ever seeing an error message, but by following its advice, you can see fewer of them.

Working with Error Messages

Error messages are designed to tell you that something has gone wrong. It seems like an obvious thing to say, but some people actually view error messages as the computer's way of telling them it dislikes them. Computers don't actually hate anyone — they simply like things done a certain way because they really don't understand any other way.

Error messages come in two levels:

✔ **Error:** When an actual error occurs, processing stops because the MATLAB application can't continue doing useful work. You can tell an error message because it appears in dark-red type and the system makes a sound. However, the most important aspect of an error is that the application stops running, and you need to fix the problem before it can proceed.

✔ **Warning:** When a warning condition occurs, the computer tells you about it but continues processing the application as long as doing so is safe. The message type shows up in a lighter red in this case, and you don't hear any sort of system sound. In addition, the error message is preceded by the word *Warning*. In fact, you might not know that a warning has happened until you look at the Command window. Fixing a warning is still important because warnings have a habit of becoming errors over time.

The most important thing to remember about error and warning messages is that someone created them based on what that person thought would happen, rather than what *is* happening. You sometimes see an error or warning message that doesn't quite make sense in the situation or is possibly completely wrong. Of course, first you try reacting to the error message, but if you can't find a cause for the error, it's time to start looking in other places. The error or warning message is designed to provide you with help after something has gone wrong — and sometimes the help is absolutely perfect and at other times it isn't, and you need to work a little harder to resolve the situation.

Responding to error messages

Until now, every time an error has happened, it has simply displayed in the Command window without much comment. However, you can do more about errors than simply wait for them to announce themselves. MATLAB lets you intercept errors and handle them in various ways using a special `try...` `catch` structure. Listing 13-1 shows how to create such a structure in a function. You can also find this function in the `Broken.m` file supplied with the downloadable source code.

Listing 13-1:   Using the try...catch Structure

```
function [ ] = Broken( )
%BROKEN A broken piece of code.
%    This example is designed to generate an error.

    try
        Handle = fopen('DoesNotExist.txt');
        Data = fread(Handle);
        disp(Data);
    catch exc
        disp('Oops, an error has occurred!');
        disp(exc)
    end
end
```

First, look at the source of the error. The call to `fopen()` uses a file that doesn't exist. Using this call isn't a problem; in fact, some calls to `fopen()` are expected to fail. When this failure happens, `fopen()` returns a handle with a value of -1. The problem with this code occurs in the *next* call. Because `Handle` doesn't contain a valid handle, the `fread()` call fails. Reading data from a file that doesn't exist isn't possible.

The `try...catch` block contains the code that you want to execute between `try` and `catch`. When an exception does occur, the information is placed in `exc`, where you can use it in whatever way is necessary. In this case, the error-handling code (*error handler*) — the code between `catch` and `end` — displays a human-readable message and a more specific MATLAB error message. The first is meant for the user; the second is meant for the developer.

To try this code, type **Broken()** and press Enter in the Command window. You see the following output:

```
Oops, an error has occurred!
  MException with properties:

    identifier: 'MATLAB:FileIO:InvalidFid'
       message: 'Invalid file identifier.  Use fopen to
                 generate a valid file identifier.'
         cause: {0x1 cell}
         stack: [1x1 struct]
```

The exception information starts with the second line. It tells you that the exception is a member of the MException class and has certain properties. Here's the additional information you receive:

- ✔ identifier: A short, specific description of the error. An identifier provides a category of errors, and you can use it to find additional information about the error as a whole.

- ✔ message: A longer message that provides details about the problem in this particular instance. The message is generally easier for humans to understand than is the other information.

- ✔ cause: When establishing a cause for the problem is possible, this property contains a list of causal sources.

- ✔ stack: The path that the application has followed to get to this point. By tracing the application path, you can often find a source for an error in some other function — the caller of the current function (or one of its callers all the way up to the main application in the call hierarchy).

Understanding the MException class

The MException class is the basis for the exception information that you receive from MATLAB. It provides you with the properties and functions needed to work with exceptions; you'd then use the exceptions to overcome, or at least reduce, the effect of errors on the user. The previous section of the chapter acquaints you with the four properties that the MException class provides. Here's a list of the functions that you use most often:

- ✔ addCause(): Appends the cause that you provide to the list of causes already provided with the exception. You can use this feature to provide ancillary information about an exception.

- ✔ getReport(): Outputs a formatted report of the exception information in a form that matches the output that MATLAB provides.

- ✔ last(): Obtains the last exception that the application threw.

- ✔ rethrow(): Sends an exception to the next higher level of the application hierarchy when handling the error at the current level isn't possible. (After an exception is accepted by an application, it's no longer active, and you must rethrow it to get another part of the application to act on it.)

- ✔ throw(): Creates an exception that either displays an error message or is handled by another part of the application.

✔ throwAsCaller(): Creates an exception using the caller's identifier that either displays an error message or is handled by another part of the application. (When one part of an application accesses a function, the part that performs the access is named the caller, so this function makes the exception appear as it if were created by the caller rather than the current function.)

One of the more interesting functions provided by the MException class is getReport(). Listing 13-2 shows how to use this function to obtain formatted output for your application. You can also find this function in the Broken2.m file supplied with the downloadable source code.

Listing 13-2:   Creating an Error Report

```
function [ ] = Broken2( )
%BROKEN2 A broken piece of code.
%    This example is designed to generate an error
%    and display a report about it.

    try
        Handle = fopen('DoesNotExist.txt');
        Data = fread(Handle);
        disp(Data);
    catch exc
        disp('Oops, an error has occurred!');
        disp(exc.getReport());
    end
end
```

Notice that you must still use the disp() function to actually output the formatted string to screen. The getReport() output is nicely formatted, if rather inflexible, because you don't have access to the individual MException properties. However, the output works fine for most uses. Here's the output of this example:

```
Oops, an error has occurred!
Error using fread
Invalid file identifier.  Use fopen to generate a valid
        file identifier.

Error in Broken2 (line 8)
                Data = fread(Handle);
```

The output includes all four properties. However, the cause isn't used in this case because the cause can't be identified by MATLAB. You need to use the addCause() function to add a cause when desired.

Creating error and warning messages

As previously mentioned, MATLAB supports both error and warning messages. You have a number of ways to create exceptions based on application conditions. The easiest way is to use the error() and warning() functions. The first creates an error condition, while the second creates a lesser, warning condition.

The example shown in Listing 13-3 presents a basic method of issuing an error or warning due to user input. However, you can use the same technique whenever an error or warning condition arises and you can't handle it locally. You can also find this function in the ErrorAndWarning.m file supplied with the downloadable source code.

Listing 13-3:   Creating Errors and Warnings

```
function [ ] = ErrorAndWarning( )
%ERRORANDWARNING Create Error and Warning Messages
%    This example shows how to create error and warning
            messages.

    NotDone = true;

    while NotDone
        try

            Value = input('Type something: ', 's');

            switch Value
                case 'error'
                    error('Input Error');
                case 'warning'
                    warning('Input Warning');
                case 'done'
                    NotDone = false;
                otherwise
                    disp(['You typed: ', Value]);
            end
        catch Exception
            disp('An exception occurred!');
            disp(Exception.getReport());
        end
    end

end
```

The example begins by creating a loop. It then asks the user to type something. If that something happens to be **error** or **warning**, the appropriate error or warning message is issued. When the user types **done**, the application exits. Otherwise, the user sees a simple output message. The example looks simple, and it is, but it has a couple of interesting features. The following steps help you work with the example:

1. **Type** ErrorAndWarning() **and press Enter in the Command window.**

   The application asks you to type something.

2. **Type** Hello World! **and press Enter.**

   You see the following output:

   ```
   You typed: Hello World!
   ```

   The application asks the user to type something else.

3. **Type** warning **and press Enter.**

   You see the following output:

   ```
   Warning: Input Warning
   > In ErrorAndWarning at 16
   ```

   Notice that the message doesn't say anything about an exception. A warning is simply an indicator that something could be wrong, not that something is wrong. As a result, you see the message, but the application doesn't actually generate an exception. The application asks the user to type something else.

4. **Type** error **and press Enter.**

   You see the following output:

   ```
   An exception occurred!
   Error using ErrorAndWarning (line 14)
   Input Error
   ```

   This time, an exception is generated. If the exception handler weren't in place, the application would end at this point. However, because an exception handler is in place, the application can ask the user to type something else. Adding exception handlers makes recovering from exceptions possible, as happens in this case. Of course, your exception handler must actually fix the problem that caused the exception.

5. **Type** done **and press Enter.**

   The application ends.

The example application uses the simple form of the error() and warning() functions. Both the error() and warning() functions can accept an identifier as the first argument, followed by the message as the second. You can also add the cause and stack trace elements as arguments. The point is, all you really need in most cases is a simple message.

Setting warning message modes

Error messages always provide you with any details that the application provides. Warning messages are different — you can tell MATLAB to provide only certain details as needed. You can set the configuration globally or base it on the message identifier (so that warnings with some identifiers provide more information than others do). To set the global configuration, you provide two arguments. The first is the warning state:

- ✔ on: Sets the configuration element on.
- ✔ off: Turns the configuration element off.
- ✔ query: Determines whether the configuration element is currently on or off.

At a global level, you have access to two settings, which are provided as the second argument to the warning() function:

- ✔ backtrace: Determines whether the output includes stack trace information.
- ✔ verbose: Determines whether the output includes a short message or one with all the details as well.

To see how this works, type **warning('query', 'backtrace')** and press Enter. You should see an output message telling you the current status of the backtrace configuration element. (The default setting turns it on so that you can see the stack trace as part of the warning message when a stack trace is provided.)

The message identifier-specific form of the warning() function starts with a state. However, in this case, the second argument is a message identifier. For example, if you type **warning('query', 'MATLAB:FileIO:InvalidFid')** and press Enter, you see the current state of the MATLAB:FileIO:InvalidFid identifier.

Setting warnings for particular message identifiers to off is usually a bad idea because the system can't inform you about problems. This is especially true for MATLAB-specific messages (rather than custom messages that you create, as described in the "Making Your Own Error Messages" section later in the chapter). However, setting them to off during troubleshooting can help you avoid the headache of seeing the message all the time.

Understanding Quick Alerts

Errors can happen at any time. In fact, there seems to be an unwritten law that errors must happen at the most inconvenient time possible and only when anyone who can possibly help is out of the building. When you face this problem, at least you have the assurance that every other person to ever write an application has faced the same problem. Waiting by your phone for a call that you don't want to hear (because it's always a bad news call) is one solution to the problem, but probably not the best solution because the person on the other end of the line is unlikely to have the information you so desperately need about the issue.

This is a situation in which sending yourself a note is probably a better option than waiting for the call. Fortunately, MATLAB provides the sendmail() function for this purpose. It's possible for you to make one of the responses in your error-handling code be sending an email that you can pick up on your smartphone. The result is that you get information directly from the application that helps you fix the problem right where you are, rather than have to go into work. The sendmail() function accepts these arguments:

- ✔ **Recipients:** Provides a list of one or more recipients for the message. Each entry is separated from the other with a semicolon.

- ✔ **Subject:** Contains the message topic. If the problem is short enough, you can actually present the error message in the subject line.

- ✔ **Message (optional):** Details the error information.

- ✔ **Attachments (optional):** Specifies the path and full filename of any attachment you want to send with the message.

Before you can send an email, you must configure MATLAB to recognize your Simple Mail Transfer Protocol (SMTP) server and provide a From address. In order to do this, you must use a special setpref() function. For example, if your server is smtp.mycompany.com, then you type **setpref('Internet', 'SMTP_Server', 'smtp.mycompany.com')** in the Command window and press Enter. After you set the SMTP address, you set the From address by providing your email address as input to the setpref() function, as in setpref('Internet', 'E_mail', 'myaddress@mycompany.com').

Listing 13-4 shows a technique for sending an email. The code used to create the error is similar to the Broken example used earlier in the chapter. However, this time the example outputs an email message rather than a message onscreen. You can also find this function in the `Broken3.m` file supplied with the downloadable source code.

Listing 13-4: Sending an Email Alert

```
function [ ] = Broken3( )
%BROKEN3 A broken piece of code.
%   This example is designed to generate an error
%   and send an e-mail about it.

    try
        Handle = fopen('DoesNotExist.txt');
        Data = fread(Handle);
        disp(Data);
    catch exc
        disp('Sending for help!');
        sendmail('myaddress@mycompany.com',...
            'Broken2',...
            ['Identifier: ', exc.identifier,10,...
            'Message: ', exc.message]);
    end
end
```

Notice how the example uses the `sendmail()` function. The address and subject appear much as you might think from creating any other email message. The message is a concatenation of strings. The first line is the error identifier and the second is the error message. Notice the number 10 between the two lines. This value actually creates a new line so that the information appears on separate lines in the output message. Figure 13-1 shows a typical example of the message (displayed using Outlook in this case).

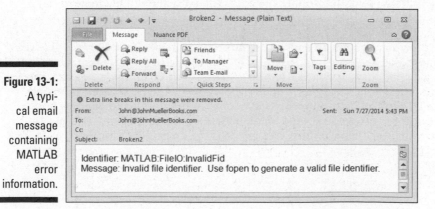

Figure 13-1: A typical email message containing MATLAB error information.

Relying on Common Fixes for MATLAB's Error Messages

MATLAB does try to inform you about errors whenever it finds them. Of course, your first line of defense in fixing those errors is to rely on MATLAB itself. Previous chapters of the book have outlined a number of these automatic fixes. For example, when you make a coding error, MATLAB usually asks whether you meant to use some alternative form of the command, and it's right quite often about the choice it provides.

The editor also highlights potential errors in orange. When you hover the mouse over these errors, you see a small dialog box telling you about the problem and offering to fix it for you. The article at `http://www.mathworks.com/help/matlab/matlab_prog/check-code-for-errors-and-warnings.html` discusses a number of other kinds of automatic fixes that you should consider using.

Sometimes an automatic fix doesn't make sense, but the combination of the error message and the automatic fix provides you with enough information to fix the problem on your own. The most important things to do are to read both the error message and the fix carefully. Humans and computers speak different languages, so there is a lot of room for misunderstanding. After you read the information carefully, look for typos or missing information. For example, MATLAB does understand `A*(B+C)` but doesn't understand `A(B+C)`. Even though a human would know that the `A` should be multiplied by the result of `B+C`, MATLAB can't make that determination. Small bits of missing text have a big impact on your application, as do seemingly small typos, such as typing `Vara` instead of `VarA`.

Don't give up immediately, but at some point you need to start consulting other resources rather than getting bogged down with an error that you can't quite fix. The MATLAB documentation can also be a good source of help, but knowing where to look (when you can barely voice the question) is a problem. That's where MATLAB Answers (`http://www.mathworks.com/matlabcentral/answers/`) comes into play. You can use this resource to obtain answers from MATLAB professionals, in addition to the usual peer support. If you can't find someone to help you on MATLAB Answers, you can usually get script and function help on Code Project (`http://www.codeproject.com/script/Answers/List.aspx?tab=active&tags=922`) and other third-party answer sites.

Fortunately, there are other documentation alternatives when the MATLAB documentation can't help. For example, the MATLAB Programming/Error Messages article at `http://en.wikibooks.org/wiki/MATLAB_Programming/Error_Messages` describes a number of common errors and how to fix them. Another good place to look for helpful fixes to common problems is

MATLAB Tips (http://www3.nd.edu/~nancy/Math20550/Matlab/tips/matlab_tips.html). In short, you have many good places to look online for third-party assistance if your first line of help fails.

Although you can find a lot of MATLAB information online, be aware that not all of it is current. Old information may work fine with a previous version of MATLAB, but it may not work at all well with the version installed on your system. When looking for help online, make sure that the help you obtain is for the version of MATLAB that you actually have installed on your machine, or test the solution with extreme care to ensure that it does work.

Making Your Own Error Messages

At some point, the standard error messages that MATLAB provides will fall short, and you'll want to create custom error messages of your own. For example, MATLAB has no idea how your custom function is put together, so the standard messages can't accommodate a situation in which a user needs to provide a specific kind of input. The only way you can tell someone that the input is incorrect is to provide a custom error message.

Fortunately, MATLAB provides the means for creating custom error messages. The following sections describe how to create the custom error messages first, and then how to ensure that your custom error messages are as useful as possible. The most important task of an error message is to inform others about the problem at hand in a manner that allows them to fix it. So creating custom error messages that really do communicate well is essential.

Developing the custom error message

This is the first place in the book that involves you in an example that's a little more complex. When you develop something like custom error messages, you want to create the code itself, followed by a means to test that code. Developers use a fancy term, *testing harness,* to describe the code used to test other code. It really isn't necessary to give it an odd name. One file contains the code that you use to check for a condition and another contains the test code. The following sections describe the two files used for this example.

Creating the exception code

Testing user inputs is usually a good idea because you never know what a user will provide. In Listing 13-5, the code performs a typical set of checks to ensure that the input is completely correct before using it to perform a task — in this case, displaying the value onscreen. The technique used in this example is a good way to ensure that no one can send you data that isn't appropriate for a particular application need. You can also find this function in the CustomException.m file supplied with the downloadable source code.

Listing 13-5:   Checking for Exceptional Conditions

```
function [ ] = CustomException( Value )
%CUSTOMEXCEPTION Demonstrates custom exceptions.
%   This example shows how to put a custom exception
%   together.

    if nargin < 1
        NoInput = MException('MyCompany:NoInput',...
            'Not enough input arguments!');
        NoInput.throw();
    end

    if not(isnumeric(Value))
        NotNumeric = MException('MyCompany:NotNumeric',...
            'Input argument is type %s number needed!',...
            class(Value));
        NotNumeric.throw();
    end

    if (Value < 1) || (Value > 10)
        NotInRange = MException('MyCompany:NotInRange',...
            'Input argument not between %d and %d!',...
            1, 10);
        NotInRange.throw();
    end

    fprintf('Successfully entered the value: %d.\r',...
        Value);

end
```

The code begins by checking the number of arguments. This example contains no default value, so not supplying a value is an error, and the code tells the caller about it. The NoInput variable contains an MException object that has an identifier for MyCompany:NoInput. This is a custom identifier, the sort you should use when creating your own exceptions. An identifier is a string, such as your company name, separated by a colon from the exception type, which is NoInput in this case.

The message following the identifier provides additional detail. It spells out that the input doesn't provide enough arguments in this case. If you wanted, you could provide additional information, such as the actual input requirements for the application.

After `NoInput` is created, the code uses the `throw()` method to cause an exception. If the caller hasn't placed the function call in a `try...catch` block, the exception causes the application to fail. The exception does cause the current code to end right after the call to `throw()`.

The second exception works much the same as the first. In this case, the code checks to ensure that the input argument (now that it knows there is one) is a numeric value. If `Value` isn't numeric, another exception is thrown. However, notice that this exception detects the kind of input actually provided and returns it as part of the message. The messages you create can use the same placeholders, such as `%d` and `%s`, as the `sprintf()` and `fprintf()` functions used in earlier chapters.

Note the order of the exceptions. The code tests to ensure that there is an argument before it tests the argument type. The order in which you test for conditions that will stop the application from running properly is essential. Each step of testing should build on the step before it.

The third exception tests the range of the input number (now that the code knows that it is indeed a number). When the range is outside the required range, the code throws an exception.

When everything works as it should, the code ends by displaying `Value`. In this case, the application uses `fprintf()` to make displaying the information easier than it would be when using `disp()`, because `disp()` can't handle numeric input.

Creating the testing code

Testing your code before using it in a full-fledged application is essential. This is especially true for error-checking code, such as that found in `CustomException()`, because you rely on such code to tell you when other errors occur. Any code that generates exceptions based on errant input must be held to a higher standard of testing, which is why you need to create the testing harness shown in Listing 13-6. You can also find this function in the `TestCustomException.m` file supplied with the downloadable source code.

Listing 13-6: Testing the Exception Code

```
function [ ] = TestCustomException( )
%TESTCUSTOMEXCEPTION Tests the CustomException() function.
%    Performs detailed testing of the CustomException()
            function
%    by checking for input type and ranges.
```

```
    % Check for no input.
    try
        disp('Testing no input.');
        CustomException();
    catch Exc
        disp(Exc.getReport());
    end

    % Check for logical input.
    try
        disp('Testing logical input.');
        CustomException(true);
    catch Exc
        disp(Exc.getReport());
    end

    % Check for string input.
    try
        disp('Testing string input.');
        CustomException('Hello');
    catch Exc
        disp(Exc.getReport());
    end

    % Check for number out of range.
    try
        disp('Testing input too low.');
        CustomException(-1);
    catch Exc
        disp(Exc.getReport());
    end
    try
        disp('Testing input too high.');
        CustomException(12);
    catch Exc
        disp(Exc.getReport());
    end

    % Check for good input.
    try
        disp('Testing input just right.');
        CustomException(5);
    catch Exc
        disp(Exc.getReport());
    end
end
```

This code purposely creates exceptions and then outputs the messages generated. By running this code, you can ensure that CustomException() works precisely as you thought it would. Notice that the test order follows the same logical progression as the code in the `CustomException.m` file. Each test step builds on the one before it. Here's the output you see when you run `TestCustomException()`:

```
Testing no input.
Error using CustomException (line 9)
Not enough input arguments!
Error in TestCustomException (line 9)
        CustomException();
Testing logical input.
Error using CustomException (line 16)
Input argument is type logical number needed!
Error in TestCustomException (line 17)
        CustomException(true);
Testing string input.
Error using CustomException (line 16)
Input argument is type char number needed!
Error in TestCustomException (line 25)
        CustomException('Hello');
Testing input too low.
Error using CustomException (line 23)
Input argument not between 1 and 10!
Error in TestCustomException (line 33)
        CustomException(-1);
Testing input too high.
Error using CustomException (line 23)
Input argument not between 1 and 10!
Error in TestCustomException (line 39)
        CustomException(12);
Testing input just right.
Successfully entered the value: 5.
```

The output shows that each phase of testing ends precisely as it should. Only the final output provides the desired result. Notice how the incorrect input types generate a custom output message that defines how the input is incorrect.

Creating useful error messages

Creating useful error messages can be hard. When you find yourself scratching your head, trying to figure out just what's wrong with your input, you're experiencing a communication problem. The error message doesn't provide enough information in the right form to communicate the problem to you. However, creating good error messages really is an art, and it takes a bit of practice. Here are some tips to make your error-message writing easier:

✔ Keep your messages as short as possible because long messages tend to become difficult to understand.

✔ Focus on the problem at hand, rather than what you think the problem might be. For example, if the error message says that the file is missing, focus on the missing file, rather than on something like a broken network connection. It's more likely that the user mistyped the filename than it is that the network is down. If it turns out that the filename really is correct, then it could have gotten erased on disk. You do need to eventually check out the network connection, but focus on the problem at hand first and then move out from there so that your error trapping is both procedural and logical.

✔ Provide specific information whenever possible by returning the errant information as part of the error message.

✔ Ask others to try your test harness, read the messages, and provide feedback.

✔ Make the error message a more detailed version of the message identifier and ensure that the message identifier is unique.

✔ Verify that every message is unique so that users don't see the same message for different conditions. If you can't create unique wording, perhaps you need to create a more flexible version of the message that works for both situations.

✔ Ensure that each message is formatted in a similar way so that users can focus on the content rather than the format.

✔ Avoid humorous or irritating language in your messages — make sure that you focus on simple language that everyone will understand and that won't tend to cause upset rather than be helpful.

Using Good Coding Practices

A lot of places online tell you about good coding practice. In fact, if you ask five developers about their five best coding practices, you get five different answers, partly because everyone is different. The following list represents the best coding practices from a number of sources (including personal experience) that have stood the test of time.

✔ **Get a second opinion:** Many developers are afraid to admit that they make mistakes, so they keep looking at the same script or function until they're bleary eyed, and usually end up making more mistakes as a result. Having someone else look at the problem could save you time and effort, and will most definitely help you discover new coding practices more quickly than if you go it alone.

✔ **Write applications for humans and not machines:** As people spend more time writing code, they start to enjoy what they do and start engaging in writing code that looks really cool but is truly horrible to maintain. In addition, the code is buggy and not very friendly to the people using it. Humans use applications. No one uses cool code — people use applications that are nearly invisible that help them get work done quickly and without a lot of fuss.

✔ **Test often/make small changes:** A few people actually try to write an entire application without ever testing it, even once, and then they're surprised when it doesn't work. The best application developers work carefully and test often. Making small changes means that you can find errors faster and fix them faster still. When you write a whole bunch of code without testing it, you really don't have any way to know where to start looking for problems.

✔ **Don't reinvent the wheel:** Take the opportunity to use someone else's fully tested and bug-free code whenever you can (as long as you don't have to steal the code to do so). In fact, actively look for opportunities to reuse code. Using code that already works in another application saves you time writing your application.

✔ **Modularize your application:** Writing and debugging a piece of coding takes time and effort. Maintaining that code takes even longer. If you have to make the same changes in a whole bunch of places every time you discover a problem with your code, you waste time and energy that you could use to do something else. Write the code just one time, place it in a function, and then access that piece of code everywhere you need it.

✔ **Plan for mistakes:** Make sure your code contains plenty of error trapping. It's easier to catch a mistake and allow your application to fail gracefully than it is to have the application crash and lose data that you must recover at some later time. When you do add error-trapping code, make sure to write it in such a manner to actually trap the sorts of errors that you expect, and then add some general-purpose error trapping for the mistakes you didn't expect.

✔ **Create documentation for your application:** Every application requires documentation. Even if you plan to use the application to meet just your own needs, you need documentation because all developers eventually forget how their code works. Professionals know from experience that good documentation is essential. When you do create the documentation, make sure that you discuss why you designed the software in a certain manner, what you were trying to achieve by creating it, problems you encountered making the software work, and fixes you employed in the past. In some cases, you want to also document how something works, but keep the documentation of code mechanics (how it works) to a minimum.

✔ **Ensure that you include documentation within your application as comments:** Comments within applications help at several different levels, the most important of which is jogging your memory when you try to figure out how the application works. However, it's also important to remember that typing help('ApplicationName') and pressing Enter will display the comments as help information to people using your application.

✔ **Code for performance after you make the application work:** Performance consists of three elements: reliability, security, and speed. A reliable application works consistently and checks for errors before committing changes to data. A secure application keeps data safe and ensures that user mistakes are caught. A fast application performs tasks quickly. However, before you can do any of these things, the application has to work in the first place. (Remember that you can use the profile() command to measure application performance and determine whether changes you implement actually work as intended.)

✔ **Make the application invisible:** If a user has to spend a lot of time acknowledging the presence of your application, your application will eventually end up collecting dust. For example, the most annoying application in the world is the one that displays those "Are you sure?" messages. If the user wasn't sure, then there would be no reason to perform the act. Instead, make a backup of the change automatically so that the user can reverse the change later. Users don't even want to see your application — it should be invisible for the most part. When a user can focus on the task at hand, your application becomes a favorite tool and garners support for things like upgrades later.

✔ **Force the computer to do as much work as possible:** Anytime you can make something easier for the user, you reduce the chance that the user will make a mistake that you hadn't considered as part of the application design. Easy doesn't mean fancy. For example, applications that try to guess what the user is going to type next usually get it wrong and only end up annoying everyone. Let the user type whatever is needed, but then you can check the input for typos and ensure that the input won't cause the application to fail. In fact, rather than try to guess what the user will type next, create an interface that doesn't actually require any typing. For example, instead of asking the user to enter a city and state in a form, have the user type a zip code and obtain the city and state from the zip code information.

Part V
Specific MATLAB Applications

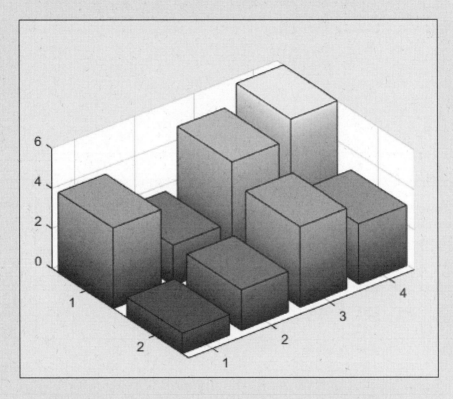

In this part . . .

- ✔ Use algebra to solve equations and find roots.
- ✔ Obtain the statistics needed for a presentation.
- ✔ Perform analysis using algebra and calculus.
- ✔ Solve differential equations.
- ✔ Create unusual plots and dress up existing plots.

Chapter 14

Solving Equations and Finding Roots

MATLAB is amazing when it comes to helping you solve equations and find roots. Of course, getting the right answer happens only when you know how to ask the right question. Communicating with MATLAB in a manner it understands is an important part of solving the questions you have. This chapter demonstrates how to solve specific kinds of equations and how to find roots. The important thing to consider as you read is how the information creates patterns that you can use to solve your specific algebraic or statistical problem.

In most cases, there are multiple ways to obtain an answer to any question. This chapter demonstrates one method for each kind of equation or root. However, you can find additional solutions online in locations such as MATLAB Answers (http://www.mathworks.com/matlabcentral/answers/). The point is that MATLAB can provide an answer as long as you have a viable means to ask the question.

Working with the Symbolic Math Toolbox

The Symbolic Math Toolbox immensely reduces the work required to solve equations. In fact, it might almost seem like magic to some people. Of course, no magic is involved — the clever programmers at MathWorks just make it look that way.

However, before you can begin using the Symbolic Math Toolbox to perform amazing feats, you need to have it installed. If you have the student version, the Toolbox is installed by default and you can skip the first two sections that follow (going right to the "Working with the GUI" section). Otherwise, start with the first section that follows to get your copy of the Symbolic Math Toolbox and install it on your system.

Obtaining your copy of the Toolbox

You need to obtain either a trial version or a purchased version of the Symbolic Math Toolbox before you can do anything else. (When getting a trial version, you must discuss the download with someone from MATLAB before you can actually download the product.) Check out the product information at `http://www.mathworks.com/products/symbolic/` and click one of the links in the Try or Buy section of the page (normally on the right side). After you have received confirmation of your purchase, use the following steps to obtain the software:

1. **Navigate to** `http://www.mathworks.com/downloads/web_downloads/` **using your browser.**

   You see the MathWorks Downloads page, as shown in Figure 14-1.

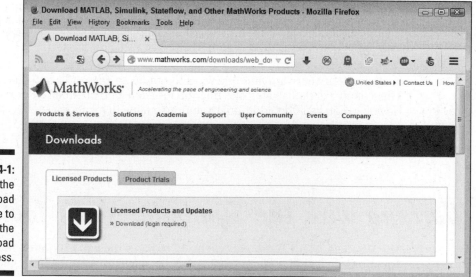

Figure 14-1:
Go to the download page to start the download process.

2. Click the Download link found on the Licensed Products tab.

You see the login page, as shown in Figure 14-2. (If you're already logged in, you don't see the login page and can go directly to Step 4.)

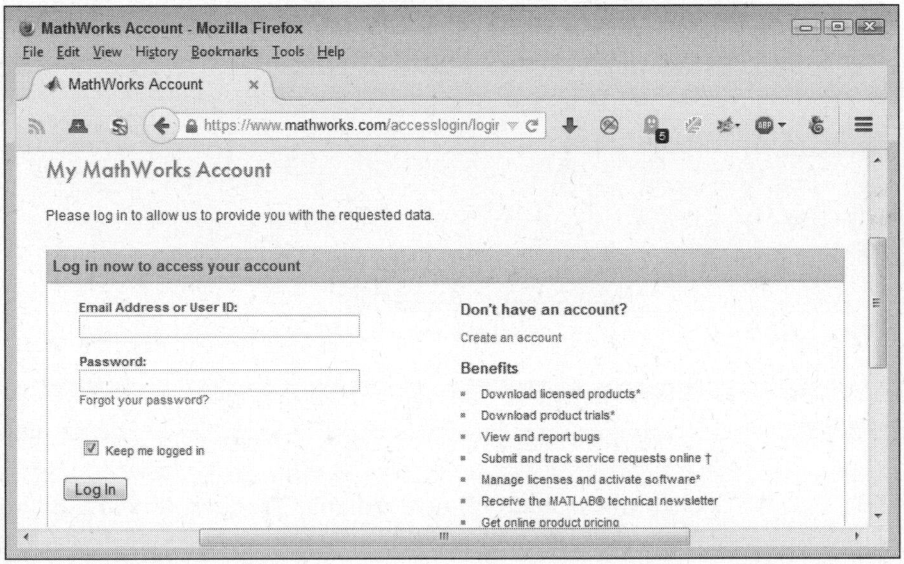

Figure 14-2:
You must login before you can download the product.

3. Type your email address and password in the appropriate fields; then click Log in.

You see a number of product releases, as shown in Figure 14-3.

4. Click the link for the MATLAB release that you have installed on your system.

When you select the product installed on your system, you should see a link for the Symbolic Math Toolbox.

5. Click the box next to the Symbolic Math Toolbox to select it.

You may need to download updates to MATLAB itself in order to install the Symbolic Math Toolbox.

6. Click Download.

Depending on your platform, you should see a dialog box for the download agent. The form this dialog box takes varies by platform. If you're able to use the download agent, the download process will start and you can skip Steps 7 and 8. Follow any directions that the download agent provides to start the download process.

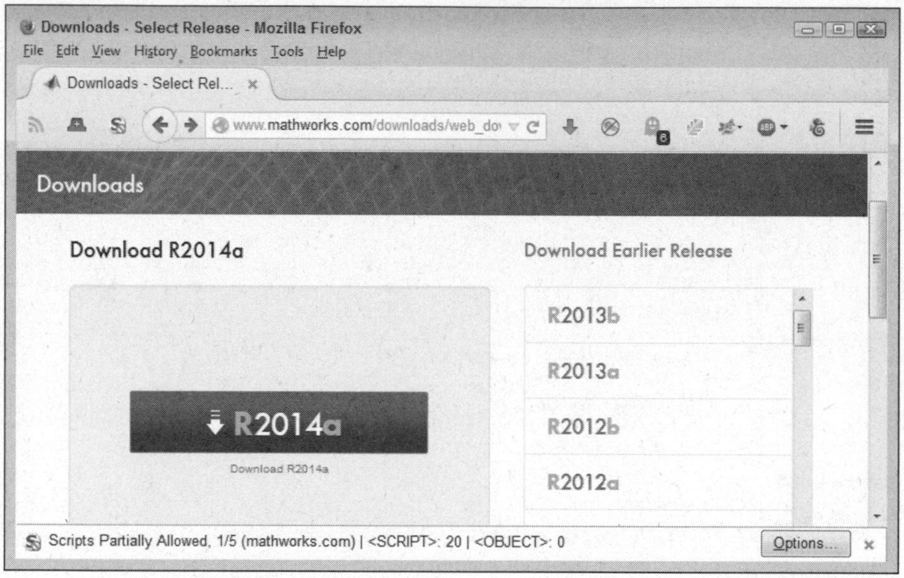

Figure 14-3:
After login,
you see a
number of
product
releases
to choose
from.

7. **Click the Manually Download Your Products link at the bottom of the page.**

   You see a download dialog box.

8. **Use the standard procedure for downloading software for your browser and platform.**

   The download is a little on the large size, so receiving it can take a while. After the download completes, you end up with a copy of the required files on your system.

Installing the Symbolic Math Toolbox

When you reach this part of the chapter, you should have a number of files on your hard drive. These files provide everything needed to install the Symbolic Math Toolbox. You have two ways by which you can interact with the files:

✔ If you were able to use the download agent, you see a dialog box telling you that the download is complete. At this point, you can perform one of these two tasks:

   • Select the Start Installer option and click Finish to start the installation process. The Symbolic Math Toolbox installer will start automatically.

- Select the Open Location of the Downloaded Files option and click Finish. You see the location of the files open, and you must double-click the installer file to start the installation process. (The installer file is typically the only executable program in the folder.)

✔ If you performed the manual download process, you need to find the download location of the files. You must double-click the installer file to start the installation process. (The installer file is typically the only executable program in the folder.)

Windows platform users may see a User Account Control (UAC) dialog box when starting the installer. Click Yes to give the installer permission to install the Symbolic Math Toolbox. Otherwise, the installation will fail.

No matter how you start the installer, eventually you see a MathWorks installer dialog box like the one shown in Figure 14-4. This dialog box determines the source of the files that you use to perform the installation. (Choosing the Install Using the Internet option downloads the files directly from the MathWorks site — you also have the option of using source files on your hard drive.) The following steps help you complete the installation process.

Figure 14-4:
The
MathWorks
Installer
dialog box
determines
the source
of files
used for the
installation
process.

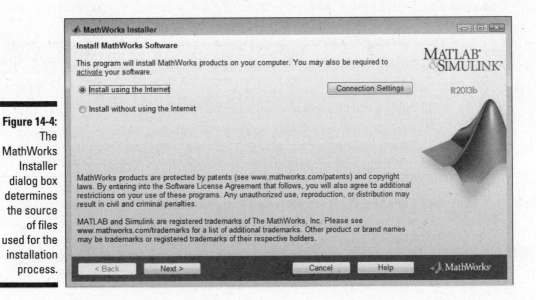

1. **Select an installation source (either Internet or local hard drive) and click Next.**

   You see the License Agreement dialog box.

2. **Read the licensing agreement, click Yes, and then click Next.**

You see the File Installation Key dialog box shown in Figure 14-5. This is where you supply the licensing information. If you don't have the key, make sure that you select the second option and follow the steps required to obtain the license.

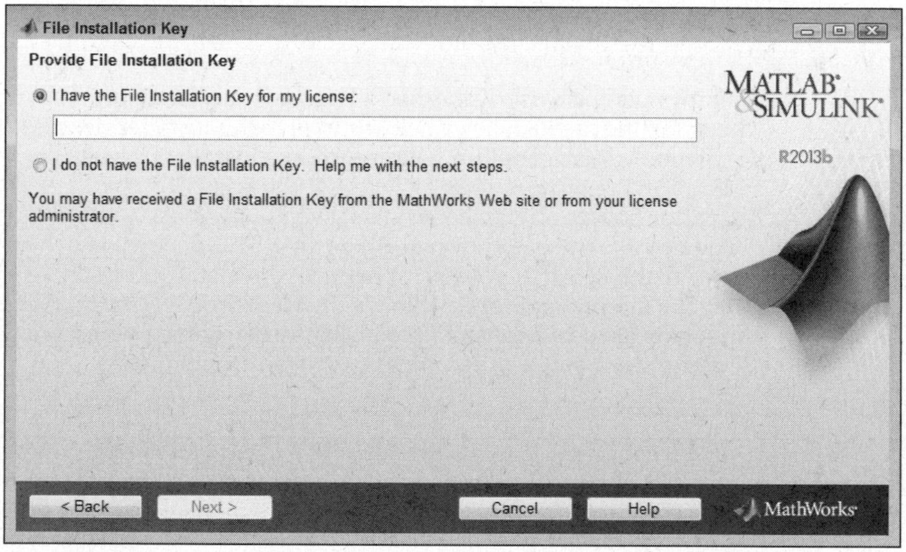

File Installation Key

Provide File Installation Key

○ I have the File Installation Key for my license:

[   ]

○ I do not have the File Installation Key. Help me with the next steps.

You may have received a File Installation Key from the MathWorks Web site or from your license administrator.

MATLAB°
&SIMULINK°

R2013b

< Back   Next >   Cancel   Help   ◢ MathWorks°

Figure 14-5:
Provide or
obtain the
file instal-
lation key
needed to
continue the
installation.

3. **Supply the File Installation Key and click Next.**

The installer asks you to select an installation method, as shown in Figure 14-6. In most cases, you obtain a better, faster, more error-free installation by selecting the Typical option. The steps that follow assume that you have chosen the Typical option.

4. **Click the Typical option and then click Next.**

The installer asks you to choose an installation destination. This destination differs by platform. In most cases, choosing the default installation destination is the best idea. However, if you have an existing installation and want to preserve this installation precisely as it is, you need to choose a different installation location.

5. **Choose a destination location, if necessary, and click Next.**

If you already have a copy of MATLAB installed and you choose the default installation location, the installer will ask whether you want to overwrite the existing copy. Click Yes To All (if you need to update your copy of MATLAB) or No (when you have the most current version) to proceed. When you click Yes To All, you agree to allow the installer to remove your old copy of MATLAB and install a new one. Be aware that you'll likely lose any special configuration options you have set up, along with any features you had installed previously.

The installer displays a Confirmation dialog box. Check the details carefully to ensure that the installation provides everything you need.

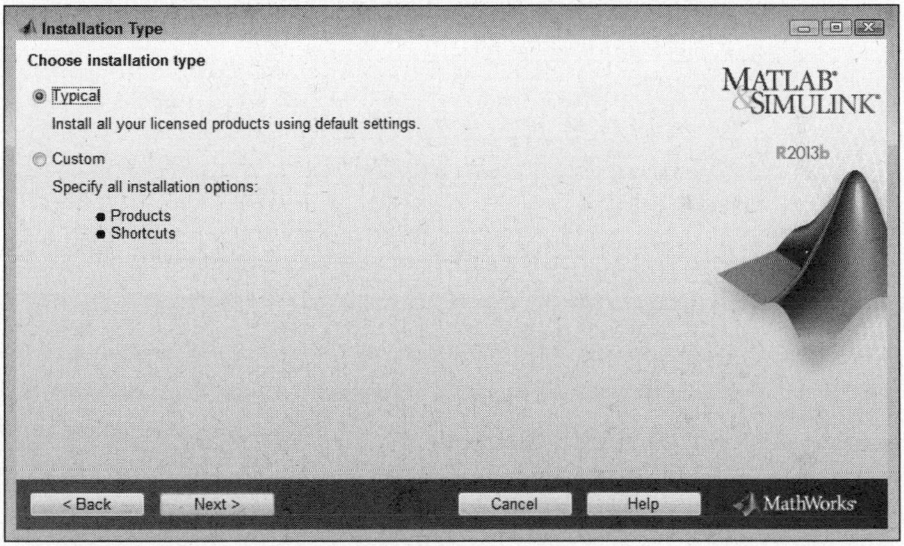

Figure 14-6:
Choose an installation method.

6. **Click Install.**

   The installation process begins. You can watch the progress by checking the progress bar. The installation can require several minutes depending on the installation options you choose, the complexity of the installation, and the speed of your system.

   When the installation process is complete, you see an Installation Complete dialog box.

7. **Select the Activate MATLAB option and then click Next.**

   MATLAB asks whether you want to activate your copy using the Internet or manually. Using the Internet is generally the faster and easier option, unless you have already downloaded a license file (`license.lic`) as part of getting the file installation key.

8. **Choose an activation option and click Next.**

   When you choose the Internet option, you must provide your email address and password to log on to the system for activation purposes. If you don't have an account, you can also choose to create an account or provide the location of your locally stored `license.lic` file, as shown in Figure 14-7.

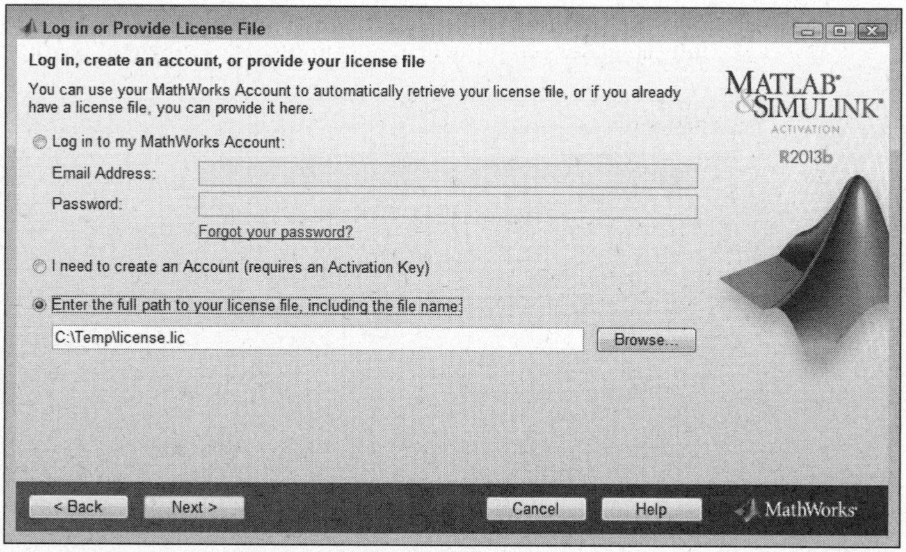

Figure 14-7:
Provide
access
to some
means of
activation.

9. **Supply any required input and click Next.**

   If activation is successful, you see an Activation Complete dialog box. (When you don't see this dialog box, retry obtaining the required activation or contact MathWorks support.)

10. **Click Finish.**

    The installation is complete.

Working with the GUI

When you install the Symbolic Math Toolbox, you see an entry added to the Apps tab — MuPAD Notebook. This is an application that you can use to create and interact with equations of various sorts. In addition, you can perform plotting based on the equations you create and work directly with matrices.

MuPAD supports a considerable number of operations; you can find the full documentation for it at `http://www.mathworks.com/discovery/mupad.html`. (This section of the chapter provides you with an extremely fast tour of MuPAD so that you get an idea of what the application can do.)

To see how MuPAD works, click the MuPAD Notepad icon on the Apps tab. You see a window like the one shown in Figure 14-8. The work area appears on the left side of the display, a toolbox (named the *Command Bar*) for performing various kinds of tasks on the right, and the usual toolbar at the top. The work area is divided into input regions (where you type a command), output regions (where you see a response), and text regions (where you can type plain text). One of the reasons to use MuPAD is to create nicely formatted reports and presentations.

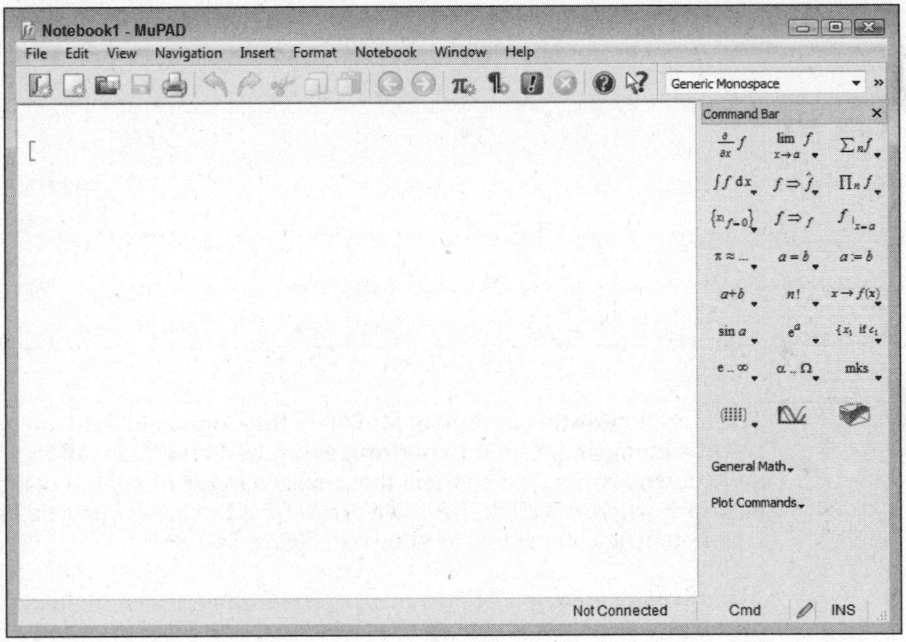

Figure 14-8: Use MuPAD to perform math tasks graphically.

You can use MuPAD to perform general math tasks. For example, when you type **1 + 1** and press Enter, you get the expected response of 2. Notice that the input region has a gray bracket next to it and that the text is in red. When you press Enter, another bracket appears directly below the first that contains the output area. This is the output region, and the text is in blue. Immediately after the first two brackets, you see another bracket. This is a new input region, and it's separated from the other two by a space, as shown in Figure 14-9.

Entering information works the same as it does with MATLAB. For example, you use the caret (^) to raise a value to a power; 7^2 is seven squared, for example. A fractional power still obtains a root. For example, 27^1/3 is the cube root of 27.

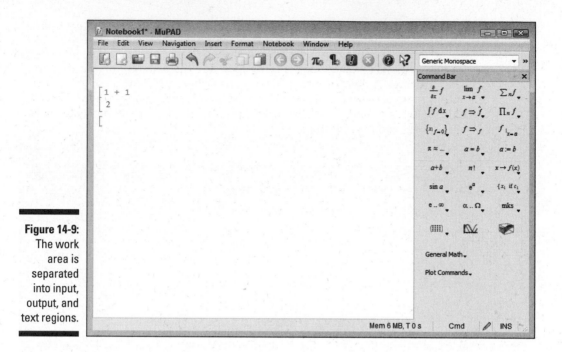

Figure 14-9:
The work area is separated into input, output, and text regions.

The more interesting feature of MuPAD is the Command Bar. You can simply select a command from it to perform a required task. Many of the entries have down arrows next to them that reveal a range of related commands. For example, when you click the down arrow next to sin a, you see a listing of trigonometric commands, as shown in Figure 14-10.

When you insert a new command, the replaceable text is highlighted. To see how this works, select n! from the Command Bar. The command actually displays as (#n)!. Type **5** and press Enter. The output region displays a value of 120. You can either type something in or select another command to place within the first one.

Unlike MATLAB, MuPAD lets you modify commands after evaluating them. Replace the 5 in the previous paragraph with a 6 and press Enter. The output region now displays a value of 720.

One of the more powerful commands is solve(). You find it in the General Math drop-down menu. solve() has a number of forms. To see one of them in action, choose General Math⇨Solve⇨Exact. The input region now displays solve(#). Type **2 * x + 3 * y - 22 = 0** and press Enter. MuPAD provides the output shown in Figure 14-11.

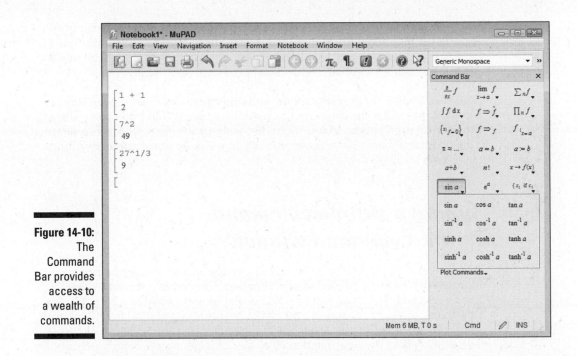

Figure 14-10:
The
Command
Bar provides
access to
a wealth of
commands.

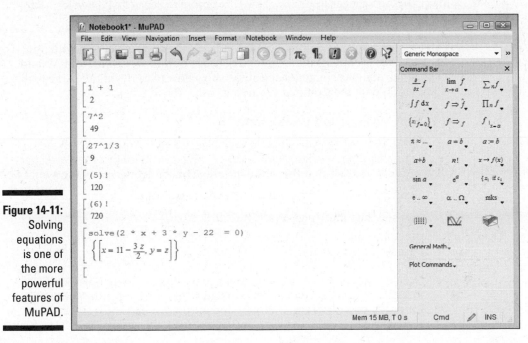

Figure 14-11:
Solving
equations
is one of
the more
powerful
features of
MuPAD.

One mistake that many people make is not remembering the requirement to include the multiplication symbol between numeric values and symbols. It's easy to type 2x instead of 2 * x. MuPAD doesn't understand that the multiplication symbol is implied, so you must provide it specifically.

MuPAD can also add graphics, plots, and other features to whatever sort of report or presentation you want to create. The idea is that you can create a nicely formatted output to use when discussing your ideas with other people. This chapter simply doesn't offer enough space to demonstrate everything (or even a small percentage of it).

Typing a simple command in the Command window

The essentials of the MATLAB Command window haven't changed since you installed the Symbolic Math Toolbox. However, you do have access to new functionality. MATLAB can access some MuPAD functionality as well as use features such as solve(). You can get all the details about MuPAD functionality at http://www.mathworks.com/help/symbolic/index.html#btgytun-1. The details about the new functions that MATLAB can access from the Symbolic Math Toolbox appear at http://www.mathworks.com/help/symbolic/functionlist.html.

The process for using solve() with MATLAB differs from using it with MuPAD. The previous section of the chapter discusses how to work with MuPAD and a specific equation. The following steps show how to perform the same task using MATLAB at the Command window:

1. **Type** syms x y **and press Enter.**

    MATLAB creates two symbolic variables, x and y. You use symbolic variables when working with solve(). When you look in the Workspace window, you see that the variables are actually defined as being symbolic.

2. **Type** solve(2 * x + 3 * y - 22 == 0) **and press Enter.**

    You see the following output:

    ```
    ans =
    11 - (3*y)/2
    ```

    Notice that you must use a double equals sign (==) when working with MATLAB, versus the single equals sign when working with MuPAD. The output is also different. Compare the output here to the output shown in Figure 14-11.

3. **Type** solve(2 * x + 3 * y - 22 == 0, y) **and press Enter.**

   This time, `solve()` solves for y rather than x. The output is now:

   ```
   ans =
   22/3 - (2*x)/3
   ```

4. **Type** solve(11 - (3*y)/2) **and press Enter.**

   You now have the value of x, which is 22/3.

5. **Type** solve(2 * 22/3 + 3 * y - 22 == 0) **and press Enter.**

   The value of y is 22/9. If you plug in both values, you find that the equation works precisely as expected. The output is now 0.

The point of these two sections is that you can use MuPAD and MATLAB in different ways with the Symbolic Math Toolbox. Both approaches are valuable, but knowing how to access the one you need is important.

Performing Algebraic Tasks

MATLAB lets you perform a wide range of algebraic tasks even without the Symbolic Math Toolbox installed, but adding the Toolbox makes performing the tasks easier. The following sections discuss using the Symbolic Math Toolbox to perform a variety of algebraic tasks. You also discover a few alternatives for performing these tasks.

Differentiating between numeric and symbolic algebra

The essential difference between numeric and symbolic algebra is that the first is used by computer science to explore principles of algebra using symbols in place of values, while the second is used by science to obtain approximations of equations for real-world use. In the "Typing a simple command in the Command window" section, earlier in this chapter, you type equations to perform symbolic algebra. In that case, you use `solve()`, which outputs a precise number (which is why you see a value of x that is 22/3). When you want to perform numeric algebra, you use the `vpasolve()` function instead. The following steps demonstrate how to perform this task:

1. **Type** syms x y **and press Enter.**

   Even when performing numeric algebra, you must define the variables before you use them.

2. **Type** vpasolve(2 * x + 3 * y - 22 == 0, x) **and press Enter.**

   You see the following output:

   ```
   ans =
   11.0 - 1.5*y
   ```

   The output is simpler this time, but notice that it also relies on floating-point numbers. In order to ensure precision, symbolic algebra relies on integers. A floating-point number is an approximation in the computer — an integer is precise.

   When working with vpasolve(), you must specify which variable to solve for. There is no assumption, and if you don't provide a variable when working with multiple variables, the output is less than useful. In this case, vpasolve() solves for x.

3. **Type** vpasolve(11.0 - 1.5*y) **and press Enter.**

   You see the following output:

   ```
   ans =
   7.3333333333333333333333333333333
   ```

   The output is a floating-point number. You aren't dealing with a fraction any longer, but the number is also an approximation. You need to note that vpasolve() defaults to providing 32-digit output — a double is 16 digits.

4. **Type** vpasolve(2 * 7.3333333333333333333333333333333 + 3 * y - 22 == 0) **and press Enter.**

   You see the following output:

   ```
   ans =
   2.4444444444444444444444444444444
   ```

   Again, the output is a floating-point number. The result is imprecise. However, seeing whether the computer can actually show you how much it's off might be interesting.

5. **Type** 2 * 7.3333333333333333333333333333333 + 3 * 2.444444444444444444444444444444444 - 22 **and press Enter.**

   MATLAB likely outputs a value of 0. The point is that the two output values truly aren't precise values, but the computer lacks the precision to detect just how much of an error exists.

Converting between symbolic and numeric data

Symbolic and numeric objects aren't compatible. You can't directly use one with the other. To make the two coexist, you must perform a conversion. Fortunately, MATLAB makes converting between symbolic and numeric data easy. The following functions perform the conversions for you:

- `double()`: Converts a symbolic matrix to a numeric form.

- `char()`: Converts symbolic objects to plain strings.

- `int8()`: Converts a symbolic matrix into 8-bit signed integers.

- `int16()`: Converts a symbolic matrix into 16-bit signed integers.

- `int32()`: Converts a symbolic matrix into 32-bit signed integers.

- `int64()`: Converts a symbolic matrix into 64-bit signed integers.

- `poly2sym()`: Converts a polynomial coefficient vector to a symbolic polynomial.

- `single()`: Converts a symbolic matrix into single-precision floating-point values.

- `sym()`: Defines new symbolic objects.

- `sym2poly()`: Converts a symbolic polynomial to a polynomial coefficient vector.

- `symfun()`: Defines new symbolic functions.

- `uint8()`: Converts a symbolic matrix into 8-bit unsigned integers.

- `uint16()`: Converts a symbolic matrix into 16-bit unsigned integers.

- `uint32()`: Converts a symbolic matrix into 32-bit unsigned integers.

- `uint64()`: Converts a symbolic matrix into 64-bit unsigned integers.

- `vpa()`: Performs a conversion between symbolic and numeric output. For example, `vpa(22/3)` results in an output of `7.333 33333333333333333333333333`.

Solving quadratic equations

There are times when using the Symbolic Math Toolbox makes things easier but using it isn't absolutely necessary. This is the case when working with quadratic equations. You actually have a number of ways to solve a quadratic equation, but two straightforward methods exist: `solve()` and `roots()`.

The `solve()` method is actually easier to understand, so type **solve(x^2 + 3*x - 4 == 0)** and press Enter. You see the following output:

```
ans =
   1
  -4
```

In this case, you work with a typical quadratic equation. The equation is entered directly as part of the `solve()` input. Of course, you need to use a double equals sign (`==`), remembering to add the multiplication operator, but otherwise, the equation looks precisely as you might write it manually.

The `roots()` approach isn't quite as easy to understand by just viewing it. Type **roots([1 3 -4])** and press Enter. As before, you get the following output:

```
ans =
    -4
     1
```

Except for being in reverse order, the outputs are the same. However, when working with `roots()`, you pass a vector containing just the constants (coefficients) for the equation in a vector. Nothing is wrong with this approach, but six months from now, you may look at the `roots()` call and not really understand what it does.

Working with cubic and other nonlinear equations

The Symbolic Math Toolbox makes it easy to solve cubic and other nonlinear equations. The example in this section explores the cubic equation, which takes the form: `ax^3+bx^2+cx+d=0`. Each of the coefficients take these forms:

- ✔ a=2
- ✔ b=-4
- ✔ c=-22
- ✔ d=24

Now that you have the parameters for the cubic equation, it's time to solve it. The following steps show you how to solve the problem in the Command window:

1. **Type** syms x **and press Enter.**

    MATLAB creates the required symbolic object.

2. **Type each of the following coefficients in turn, pressing Enter after each coefficient:**

```
a=2;
b=-4;
c=-22;
d=24;
```

3. **Type** Solutions = solve(a*x^3 + b*x^2 + c*x + d == 0) **and press Enter.**

You see the following output:

```
Solutions =
    1
    4
   -3
```

Of course, you can get far fancier than the example shown here, but the example gives you a good starting point. The main thing to consider is the coefficients you use. (If you ever want to check your answers, the Cubic Equation Calculator at http://www.1728.org/cubic.htm can help.)

Understanding interpolation

MATLAB supports a number of types of interpolation. (See http://whatis.techtarget.com/definition/extrapolation-and-interpolation for a description of interpolation.) You can see an overview of support for interpolation at http://www.mathworks.com/help/matlab/interpolation-1.html. For this section, you work with 1D interpolation using the interp1() function. The following steps show how to perform the task:

1. **Type** x = [0, 2, 4]; **and press Enter.**

2. **Type** y = [0, 2, 8]; **and press Enter.**

These two steps create a series of points to use for the interpolation.

3. **Type** x2 = [0:.1:4]; **and press Enter.**

At this point, you need to calculate the various forms of interpolation: linear, nearest, spline, and pchip. Steps 4 through 7 take you through this process. (Older versions of MATLAB also had a cubic option that's been replaced by pchip.)

4. **Type** y2linear = interp1(x, y, x2); **and press Enter.**

5. **Type** y2nearest = interp1(x, y, x2, 'nearest'); **and press Enter.**

6. **Type** y2spline = interp1(x, y, x2, 'spline'); **and press Enter.**

7. **Type** y2pchip = interp1(x, y, x2, 'pchip'); **and press Enter.**

   At this point, you need to plot each of the interpolations so that you can see them onscreen. Steps 8 through 11 take you through this process.

8. **Type** plot(x,y,'sk-') **and press Enter.**

   You see a plot of the points, which isn't really helpful, but it's the starting point of the answer.

9. **Type** hold on **and press Enter.**

   The plot will contain several more elements, and you need to put the plot into a hold state so that you can add them.

10. **Type** plot(x2, y2linear, 'g–') **and press Enter.**

    You see the interpolation added to the figure (along with the others as you plot them).

11. **Type** plot(x2, y2nearest, 'b–') **and press Enter.**

12. **Type** plot(x2, y2spline, 'r–') **and press Enter.**

13. **Type** plot(x2, y2pchip, 'm–') **and press Enter.**

14. **Type** legend('Data','Linear', 'Nearest', 'Spline', 'PCHIP', 'Location', 'West') **and press Enter.**

    You see the result of the various calculations, as shown in Figure 14-12.

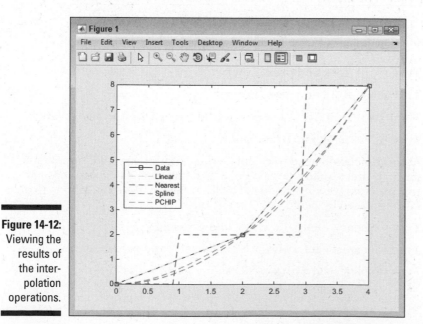

Figure 14-12:
Viewing the results of the inter-polation operations.

15. Type hold off **and press Enter.**

MATLAB removes the hold on the figure.

Working with Statistics

Statistics is an interesting area of math that deals with the collection, organization, analysis, interpretation, and presentation of data. You use it to determine the probability of the next customer buying your new widget instead of the obviously inferior widget offered by your competition. The fact is that modern business couldn't exist without the use of statistics.

MATLAB provides basic statistical support. However, if your living depends on working with statistics and you find the default MATLAB support a little lacking, you can check out the Statistics Toolbox at http://www.mathworks. com/products/statistics/ to gain additional functionality. Likewise, if you perform a lot of curve fitting, you may find that the Curve Fitting Toolbox found at http://www.mathworks.com/products/curvefitting/ comes in handy. (Admittedly, you *can* perform elementary ad hoc curve fitting in the figure window, but it's usually not sufficient to get the results you want.) The following sections don't use either the Statistics Toolbox or the Curve Fitting Toolbox for examples.

Understanding descriptive statistics

When working with *descriptive statistics,* the math quantitatively describes the characteristics of a data collection, such as the largest and smallest values, the mean value of the items, and the average. This form of statistics is commonly used to summarize the data, thus making it easier to understand. MATLAB provides a number of commands that you can use to perform basic statistics tasks. The following steps help you work through some of these tasks:

1. Type rng('shuffle', 'twister'); **and press Enter.**

You use the rng() function to initialize the pseudo-random number generator to produce a sequence of pseudo-random numbers. Older versions of MATLAB use other initialization techniques, but you should rely on the rng() function for all new applications.

The first value, shuffle, tells MATLAB to use the current time as a seed value. A *seed* value determines the starting point for a numeric sequence so that the pattern doesn't appear to repeat. If you want to exactly repeat the numeric sequence for testing purposes, you should provide a number in place of shuffle.

The second value, `twister`, is the number generator to use. MATLAB provides a number of these generators so that you can further randomize the numeric sequences you create. The upcoming "Creating pseudo-random numbers" sidebar discusses this issue in more detail.

2. **Type** w = 100 * rand(1, 100); **and press Enter.**

   This command produces 100 pseudo-random numbers that are uniformly distributed between the values 0 and 1. The numbers are then multiplied by 100 to bring them up to the integer values used in Steps 4 and 5.

3. **Type** x = 100 * randn(1, 100); **and press Enter.**

   This command produces 100 pseudo-random numbers that are normally distributed. The numbers can be positive or negative, and multiplying by 100 doesn't necessarily ensure that the numbers are between −100 and 100 (as you see later in the procedure).

4. **Type** y = randi(100, 1, 100); **and press Enter.**

   This command produces 100 pseudo-random integers that are uniformly distributed between the values of 0 and 100.

5. **Type** z = randperm(200, 100); **and press Enter.**

   This command produces 100 unique pseudo-random integers between the values of 0 and 200. There is never a repeated number in the sequence, but the 100 values are selected from the range of 0 to 200.

6. **Type** AllVals = [w; x; y; z]'; **and press Enter.**

   This command creates a 100 x 4 matrix for plotting purposes. Combining the four values lets you create a plot with all four distributions without a lot of extra steps.

7. **Type** hist(AllVals, 50); **and press Enter.**

   You see a histogram created that contains all four distributions.

8. **Type** legend('rand', 'randn', 'randi', 'randperm'); **and press Enter.**

   Adding a legend helps you identify each distribution, as shown in Figure 14-13. Notice how the various distributions differ. Only the `randn()` distribution provides both positive and negative output.

9. **Type** set(gca, 'XLim', [0, 200]); **and press Enter.**

   Figure 14-14 shows a close-up of the `rand()`, `randi()`, and `randperm()` distributions, which are a little hard to see in Figure 14-12. Notice the relatively even lines for `randperm()`. The `rand()` and `randi()` output has significant spikes.

This procedure has demonstrated a few aspects of working with statistics, the most important of which is that choosing the correct function to generate your random numbers is important. When viewing the results of your choices, you can use plots such as the histogram. In addition, don't forget that you can always modify the appearance of the plot to get a better view of what you have accomplished.

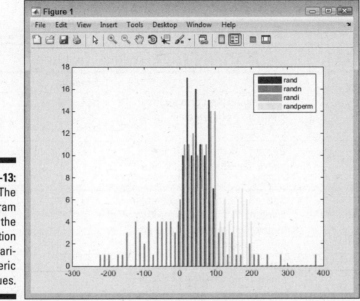

Figure 14-13:
The histogram shows the distribution of the various numeric values.

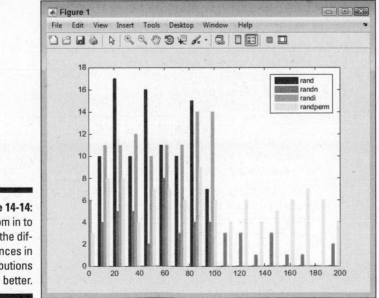

Figure 14-14:
Zoom in to see the differences in distributions better.

Of course, you can interact with the vectors in other ways. For example, you can use standard statistical functions on them. (If you have forgotten what some of these terms mean, check out `http://anothermathgeek.hubpages.com/hub/How-to-calculate-simple-statistics`.) Table 14-1 contains a list of the functions, tells what they do, and provides a short example based on the example in the steps in this section.

Table 14-1	MATLAB Basic Statistical Functions	
Function	*Usage*	*Example*
`corrcoef()`	Determines the correlation coefficients between members of a matrix.	`corrcoef(AllVals)`
`cov()`	Determines the covariance matrix for either a vector or a matrix.	`cov(AllVals)`
`max()`	Specifies the largest element in a vector. When working with a matrix, you see the largest element in each row.	`max(w)`
`mean()`	Calculates the average or mean value of a vector. When working with a matrix, you see the mean for each row.	`mean(w)`
`median()`	Calculates the median value of a vector. When working with a matrix, you see the median for each row.	`median(w)`
`min()`	Specifies the smallest element in a vector. When working with a matrix, you see the smallest element in each row.	`min(w)`
`mode()`	Determines the most frequent value in a vector. When working with a matrix, you see the most frequent value for each row.	`mode(w)`
`std()`	Calculates the standard deviation for a vector. When working with a matrix, you see the standard deviation for each row.	`std(w)`
`var()`	Determines the variance of a vector. When working with a matrix, you see the variance for each row.	`var(w)`

Creating pseudo-random numbers

Creating truly random numbers on a computer is impossible without resorting to some really exotic techniques. All random sequences on a computer are generated by algorithms, making them pseudo random. The numeric sequence has a pattern that eventually repeats itself. Depending on the algorithm used, the sequence can be quite long and, to a human, nearly indistinguishable from a random sequence. However, other computers aren't fooled, and any computer can eventually detect the repetition of the sequence.

MATLAB provides three methods of making the pseudo-random numbers created by it appear more random. The first is the seed value. Choosing a different starting point in the numeric sequence (which is quite large) means that people are less likely to actually notice any repetition. The second method provides four different distributions: `rand()` produces a uniform distribution of random numbers; `randn()` produces a normalized distribution of random numbers; `randi()` produces a uniform distribution of integers (where only whole numbers are needed); and `randperm()` produces a random permutation (where each number appears only once). The third method provides a number of randomizing generators (or engines) — essentially different algorithms — to produce the numeric sequence:

- `'combRecursive'`: Relies on the combined Multiple Recursive Generator (MRG) (see `http://random.mat.sbg. ac.at/results/karl/server/ node7.html` for details).

- `'multFibonacci'`: Uses the Multiplicative Lagged Fibonacci Generator (MLFG) variant of the Lagged Fibonacci Generator (LFG) (see `http:// northstar-www.dartmouth.edu/ doc/sprngsv1.0/DOCS/www/ paper/node13.html` for details).

- `'twister'`: Relies on the Mersenne Twister algorithm (see `http://www.math.sci. hiroshima-u.ac.jp/~m-mat/MT/ emt.html` for details).

- `'v5uniform'`: Specifies the legacy MATLAB 5.0 uniform generator, which produces the same result as the `'state'` option with `rand()`. You should use the `rng()` function to randomize the generators, rather than the `rand()` function used in the past (see `http://www. mathworks.com/help/matlab/ math/updating-your-random- number-generator-syntax.html` for details).

- `'v5normal'`: Specifies the legacy MATLAB 5.0 normal generator, which produces the same result as the `'state'` option with `randn()`. You should use the `rng()` function to randomize the generators, rather than the `randn()` function used in the past (see `http://www. mathworks.com/help/matlab/ math/updating-your-random- number-generator-syntax.html` for details).

- `'v4'`: Specifies the legacy MATLAB 4.0 generator, which produces the same result as using the `'seed'` option. This option has been replaced by newer, better, pseudo-random number generators, and you should use it only when necessary to precisely reproduce a number sequence (see `http://www. mathworks.com/help/matlab/ math/updating-your-random- number-generator-syntax.html` for details).

Understanding robust statistics

Robust statistics is a form of descriptive statistics in which the extreme values are discarded in favor of analysis with smaller changes and less potential for error. You use robust statistics when you have a potential for error in the extreme values. For example, you might use it when trying to figure out the average height of drivers today compared with those of 1940. However, you wouldn't use it when building a bridge because the extreme values are important in this second case. When working with MATLAB without any of the specialized toolkits, the best way to create robust statistics is to simply eliminate the largest and smallest values from a vector.

The easiest way to find and remove the largest and smallest values is to use the statistical functions found in Table 14-1. For example, to remove the largest value from a vector, a, you use a(a == max(a)) = [];. The max(a) part of the command finds the maximum value in vector a. The index (a == max(a)) tells MATLAB to find the index where the maximum value resides. You then set this element to an empty value, which deletes it.

Removing the smallest value from a vector is almost the same as removing the largest value. However, in this case you use a(a == min(a)) = [];. Notice that the min(a) function has taken the place of max(a).

To verify that the changes are successful, you use the std(), or standard deviation, function. As you remove large and small values, you start to see smaller std() output values.

Employing least squares fit

Least squares fit is a method of determining the best curve to fit a set of points (you can read more about this process at http://mathworld.wolfram.com/LeastSquaresFitting.html). In order to perform many of the tasks in Chapters 14 and 15, you need the Symbolic Math Toolbox. However, least squares fit is one task that you can do without the Toolbox. The following sections show both techniques. What you need to take away from these sections is that multiple methods of performing many tasks are available, so you don't absolutely have to have the Toolbox to do it, but the Toolbox does save time.

Using MATLAB alone

In order to compute this information using just MATLAB, you need to do a lot of typing. The following steps get you started. The output is the parameters and the sum of the squares of the residuals. If you want to obtain additional information, such as the 95 percent confidence level used by some people, you need to perform additional coding.

1. **Type** XSource = 1:1:10; **and press Enter.**

2. **Type** YSource = [1, 2, 3.5, 5.5, 4, 3.9, 3.7, 2, 1.9, 1.5]; **and press Enter.**

   The XSource and YSource vectors create a series of points to use for the least squares fit. The two vectors must be the same size.

3. **Type** plot(XSource, YSource) **and press Enter.**

   You see a plot of the points, as shown in Figure 14-15, which is helpful in visualizing how this process might work.

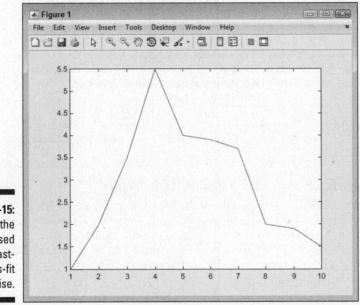

Figure 14-15: A plot of the points used for the least-squares-fit exercise.

4. **Type** fun = @(p) sum((YSource - (p(1)*cos(p(2)*XSource)+p(2)*sin(p(1)*XSource))).^2); **and press Enter.**

   This complex bit of typing is actually a function. You can use functions to automate the process of working with complex equations like this one. The equation is based on the least-squares-fitting methods described on various sites (such as the MathWorld site listed in the introduction to this section). The function accepts a single input — a guess as to the parameters for the least squares fit.

5. **Type** Guess = [2, 2]; **and press Enter.**

To make the function work, you have to provide a guess. Your guesses affect the output of the function, just as they do when performing the calculation manually.

6. **Type** [p, fminres] = fminsearch(fun, Guess) **and press Enter.**

The `fminsearch()` function accepts the function that you created and the guess that you made. Essentially, it performs unconstrained, nonlinear optimization of the function based on the guess that you provide. In this case, you see an output of

```
p =
    1.6204    1.8594
fminres =
  104.9327
```

When using this approach, you can use the output values of p for your next guess. In this case, you'd type **Guess = [1.6204, 1.8594]** and press Enter to change the guess value. Then you'd type **[p, fminres] = fminsearch(fun, Guess)** and press Enter to obtain the new output value of

```
p =
    1.6205    1.8594
fminres =
  104.9327
```

Using MATLAB with the Symbolic Math Toolbox

When working with the Symbolic Math Toolbox, you can use MuPAD to make things easier. In addition, the Symbolic Math Toolbox can greatly reduce the work you need to do by performing some of the calculations for you. The following steps assume that you have the Symbolic Math Toolbox installed and that you've worked through the basic material in the "Working with the GUI" section, earlier in this chapter.

1. **Open MuPAD by clicking the MuPAD Notebook entry on the Apps tab.**

You see a new notebook open.

2. **Type** XSource := [1, 2, 3, 4, 5, 6, 7, 8, 9,10]: **and press Enter.**

This command creates the same XSource vector as that used for the previous example. To assign the vector to XSource, you use :=, rather than just the assignment operator used in MATLAB (=). Adding the colon (:) to the end of the statement keeps MuPAD from providing output.

3. **Type** YSource := [1, 2, 3.5, 5.5, 4, 3.9, 3.7, 2, 1.9, 1.5]: **and press Enter.**

Again, this is the same set of elements used for the example in the preceding section. You now have the points needed for the least squares fit.

4. **Type** stats::reg(XSource,YSource,p1*cos(p2*x)+p2*sin(p1*x),[x],[p1,p2], StartingValues=[2, 2]) **and press Enter.**

This long statement performs the same tasks as Steps 4, 5, and 6 in the preceding example. So, even though this example looks more complex, it actually saves steps. You see the output shown in Figure 14-16.

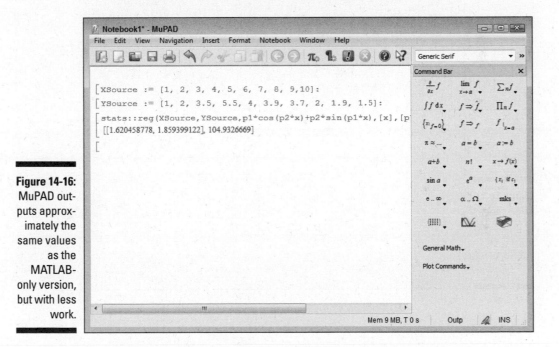

Figure 14-16: MuPAD outputs approximately the same values as the MATLAB-only version, but with less work.

The 1.620458778, 1859399122 part of the output are the parameters. You can use them to make your next guess.

5. **Highlight the 2, 2 part of the equation and type** 1.620458778, 1.859399122.

MuPAD replaces the old values with the new values you typed.

6. **Press Enter.**

You see the updated values shown in Figure 14-17. Again, they're pretty close to the values output by the MATLAB-only solution.

Using the Symbolic Math Toolbox saves time and effort by reducing the number of steps you must take to find a solution. However, the output isn't any different from working with MATLAB alone (a really good thing). The biggest time savings comes from being able to make guesses a lot faster and with greater ease.

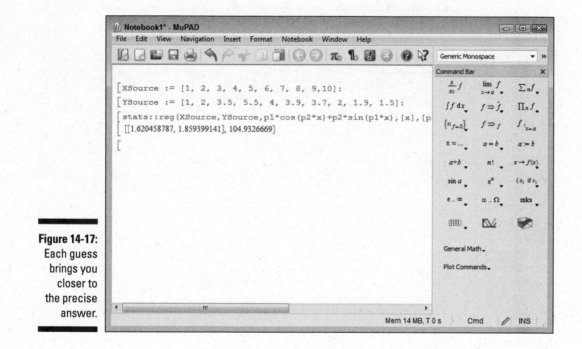

Figure 14-17:
Each guess brings you closer to the precise answer.

Chapter 15

Performing Analysis

● ●

In This Chapter

▶ Working with linear algebra

▶ Using calculus

▶ Resolving differential equations

● ●

C hapter 14 introduces you to the Symbolic Math Toolbox and shows you how to use it to perform a number of tasks. This chapter expands on some of the information presented in Chapter 14. In fact, before you proceed, make sure that you have the Symbolic Math Toolbox installed. The student version of MATLAB comes with the Symbolic Math Toolbox installed by default — otherwise, you must install it manually, using the instructions found in the first section of Chapter 14.

This chapter doesn't provide you with a detailed look at any one particular area of solving equations, but it does provide a good start on working with linear algebra, calculus, and differential equations. Being able to solve these equations quickly and correctly can make a huge difference in the success of your project.

You may think that these kinds of math are used only in high-tech environments, such as building a sub that can safely traverse the Marianas Trench. (You can read about James Cameron's successful exploration of the trench using a custom submarine at http://news.nationalgeographic. com/news/2012/03/120325-james-cameron-mariana-trench- challenger-deepest-returns-science-sub/.) However, these equations are used in everyday life. For example, check out this story about how linear algebra can be employed to make a restaurant more profitable: http://smallbusiness.chron.com/restaurants-use-linear- programming-menu-planning-37132.html. The point is that you don't really know when or where you'll encounter these equations, so it's a good idea to be prepared to use them.

Using Linear Algebra

You use linear algebra to perform a number of tasks with matrixes in MATLAB. For example, you can determine whether a matrix is singular or unimodular by using the det() function. You can also reduce a matrix to determine whether it's solvable. In fact, you can perform a relatively wide range of tasks using linear algebra with MATLAB and the Symbolic Math Toolbox; the following sections tell you how. (See Chapter 14 for details on the Symbolic Math Toolbox add-on.).

Working with determinants

Determinants are used in the analysis and solution of systems of linear equations. A nonzero value means that the matrix is nonsingular and that the system has a unique solution. A value of 1 usually indicates that the matrix is *unimodular* — that it's a real square matrix, in other words. The function used to obtain the determinant value is det(). You supply a matrix, and the output value tells you about the ability to create a solution for that matrix.

Equally important is to know about the cond() function, which tests for singular matrices. Again, you supply a matrix as an input value, and the output provides a condition number that specifies the sensitivity of the matrix to error. An output value near 1 indicates a *well-conditioned* matrix. (The YouTube video at https://www.youtube.com/watch?v=JODxbi9B3tg provides an incredibly simplified illustration of the difference between a well-conditioned and an ill-conditioned matrix.)

To see how these two functions work together, type **A = [1, 2, 3; 4, 5, 6; 7, 8, 9];** and press Enter to create a test matrix. This is a singular matrix. Type **cond(A)** and press Enter. The result of 3.8131e+16 tells you that this is a highly sensitive matrix — a singular matrix. Type **det(A)** and press Enter. Again, the incredibly small output value of 6.6613e-16 tells you that this is a singular matrix. (If you want to see a perfect singular matrix, try [0, 0, 0; 0, 0, 1; 0, 0, 0]; the cond() value is Inf, or infinity, and the det() value is 0.)

For comparison purposes, try a unimodal matrix. Type **B = [2, 3, 2; 4, 2, 3; 9, 6, 7];** and press Enter to create the matrix. Type **cond(B)** and press Enter to see the condition number of 313.1721, which isn't perfect, but it's quite close. Type **det(B)** and press Enter to see the result of 1.0000, which is good (doesn't approximate 0) for a unimodal matrix.

Performing reduction

Reduction lets you see the structure of what a matrix represents, as well as to write solutions to the system. MATLAB provides the rref() function to produce the Reduced Row Echelon Form (RREF). (You can find out more about RREF at http://www.millersville.edu/~bikenaga/linear-algebra/row-reduction/row-reduction.html.) There is also an interesting tool that you can use to see the steps required to produce RREF using any matrix as input at http://www.math.odu.edu/~bogacki/cgi-bin/lat.cgi?c=rref. The point is that you can perform reduction using MATLAB, and doing so requires only a couple of steps.

The first step is to create the matrix. In this case, the example uses a magic square. Type **A = magic(5)** and press Enter. The magic() function will produce a magic square of any size for you. (You can read about magic squares at http://mathworld.wolfram.com/MagicSquare.html). The output you see is

```
A =
    17    24     1     8    15
    23     5     7    14    16
     4     6    13    20    22
    10    12    19    21     3
    11    18    25     2     9
```

The second step is to perform the reduction. Type **rref(A)** and press Enter. Any nonsingular matrix will reduce to identity, as follows:

```
ans =
     1     0     0     0     0
     0     1     0     0     0
     0     0     1     0     0
     0     0     0     1     0
     0     0     0     0     1
```

You can use rref() to solve linear equations. In this case, if A*x=y and y=[1;0;0;0;0], then B=rref([A,y]) solves the equation. The following steps demonstrate how this works:

1. **Type** y=[1;0;0;0;0]; **and press Enter.**
2. **Type** A=magic(5); **and press Enter.**
3. **Type** B=rref([A,y]) **and press Enter.**

   You see the following output:

```
B =
    1.0000         0         0         0         0   -0.0049
         0    1.0000         0         0         0    0.0431
         0         0    1.0000         0         0   -0.0303
         0         0         0    1.0000         0    0.0047
         0         0         0         0    1.0000    0.0028
```

4. **Type** x=B(:,6) **and press Enter.**

You see the following output:

```
x =
   -0.0049
    0.0431
   -0.0303
    0.0047
    0.0028
```

At this point, you want to test the equation.

5. **Type** A*x **and press Enter.**

You see the following output:

```
ans =
    0.9999
   -0.0001
   -0.0001
   -0.0001
   -0.0001
```

Notice that the output values match the original value of y to within a small amount. In other words, the steps have proven the original equation, A*x=y, true.

Using eigenvalues

An *eigenvalue* (v) is an eigenvector of a square matrix. The variable A is a non-zero matrix. When v is multiplied by A, it yields a constant multiple of v that is commonly denoted by λ. Eigenvalues are defined by the following equation:

```
Av = λv
```

Eigenvalues are used in all sorts of ways, such as for graphics manipulation (sheer mapping) and analytic geometry (to display an arrow in three dimensional space). You can read more about eigenvalues at http://mathworld.wolfram.com/Eigenvalue.html.

To see how this works, you first need to create a matrix. Type **A = gallery('riemann', 4)** and press Enter. The gallery() function produces test matrices of specific sizes filled with specific information so that you can repeat test results as needed. The output of gallery() depends on the matrix size and the function used to create the matrix. (You can read more about gallery() at http://www.mathworks.com/help/matlab/ref/gallery.html.) The output from this particular call is

```
    1    -1     1    -1
   -1     2    -1    -1
   -1    -1     3    -1
   -1    -1    -1     4
```

Obtaining the eigenvalue comes next. The output will contain one value for each row of the matrix. Type **lambda = eig(A)** and press Enter to see the eigenvalue of the test matrix, A, as shown here:

```
lambda =
   -0.1249
    2.0000
    3.3633
    4.7616
```

Understanding factorization

Factorization is the decomposition of an object, such as a number, polynomial, or matrix. The idea behind factorization is to reduce the complexity of the object so that it's easier to understand and solve. In addition, it helps you determine how the object is put together, such as its use for prime factorization (see http://www.calculatorsoup.com/calculators/math/prime-factors.php). You can read more about factorization at http://mathworld.wolfram.com/Factorization.html.

You perform factorization in MATLAB using the factor() function. You can use the factor() function in a number of ways: working with numbers, working with polynomials, and working with matrices.

When working with a number, you simply provide the number as input. For example, type **factor(2)** and press Enter. The output is 2 because 2 is a prime number. Type **factor(12)** and press Enter. The output is [2, 2, 3] because 2 * 2 * 3 equals 12.

Polynomials require that you declare symbolic objects first by using syms. Type **syms x y** and press Enter to create the required objects. Type **factor(x^2 + 2*x*y + y^2)** and press Enter. The output is (x + y)^2.

Working with matrices requires a little more work. Begin by creating a matrix by typing **A = magic(4)** and pressing Enter. You see the following output:

```
A =
    16     2     3    13
     5    11    10     8
     9     7     6    12
     4    14    15     1
```

The factorization takes place on an element-by-element basis. However, you must enclose the matrix within the `sym()` function to get the `factor()` function to accept it. Type **factor(sym(A))** and press Enter. The output that follows shows how each element in the matrix is factorized:

```
ans =
[ 2^4,    2,    3,     13]
[   5,   11,  2*5,    2^3]
[ 3^2,    7,  2*3,  2^2*3]
[ 2^2,  2*7,  3*5,      1]
```

Employing Calculus

Calculus can solve myriad problems that algebra can't. It's really the study of how things change. This branch of math is essentially split into two pieces: differential calculus, which considers rates of change and slopes of curves, and integral calculus, which considers the accumulation of quantities and the areas between and under curves. The following sections show how you can use MATLAB with the Symbolic Math Toolbox to solve a number of relatively simple calculus problems.

Working with differential calculus

MATLAB offers good differential calculus support. The example in this section starts with something simple: Univariate differentiation. (Remember that *univariate* differentiation has a single variable.) MATLAB supports a number of forms of differential calculus — each of which requires its own set of functions. In this case, you use the `diff()` function to perform the required tasks. The following steps help you perform a simple calculation:

1. **Type** syms x **and press Enter.**

   MATLAB creates a symbolic object to use in the calculation.

2. **Type** f(x) = sin(x^3) **and press Enter.**

   Doing so creates the symbolic function used to perform the calculation. Here's the output you see:

   ```
   f(x) =
   sin(x^3)
   ```

3. **Type** Result = diff(f) **and press Enter.**

   The output shows the result of the differentiation:

   ```
   Result(x) =
   3*x^2*cos(x^3)
   ```

REMEMBER

Result(x) is actually a symbolic function. You can use it to create a picture of the output.

4. **Type** plot(Result(1:50)) **and press Enter.**

Figure 15-1 shows the plot created from the differentiation of the original symbolic function.

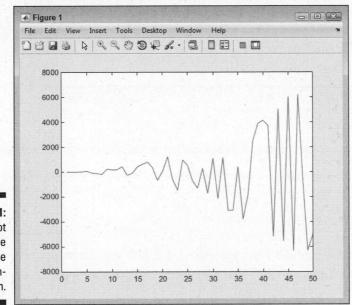

Figure 15-1:
The plot shows the result of the differentiation.

Using integral calculus

You'll also find great integral calculus support in MATLAB. As does the example in the preceding section, the example in this section focuses on a univariate calculation. In this case, the example relies on the int() function to perform the required work. The following steps help you perform a simple calculation:

1. **Type** syms x **and press Enter.**

MATLAB creates a symbolic object to use in the calculation.

2. **Type** f(x) = (x^3 + 3*x^2) / x^3 **and press Enter.**

The symbolic function that you create produces the following output:

```
f(x) =
(x^3 + 3*x^2)/x^3
```

3. **Type** Result = int(f, x) **and press Enter.**

   Notice that you must provide a symbolic variable as the second input. The output shows the following symbolic function as the result of the integration:

   ```
   Result(x) =
   x + 3*log(x)
   ```

4. **Type** plot(Result(1:50)) **and press Enter.**

   Figure 15-2 shows the plot created from the integration of the original symbolic function.

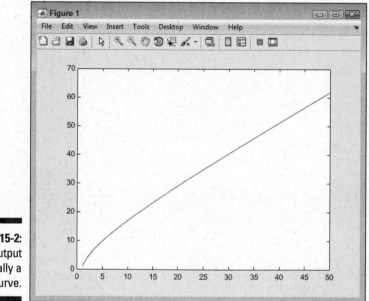

Figure 15-2:
The output is usually a curve.

Working with multivariate calculus

The "Working with differential calculus" section, earlier in the chapter, shows how to work with a single variable. Of course, many (if not most) problems don't involve just one variable. With this in mind, the following steps demonstrate a problem with more than one variable — a *multivariate* example:

1. **Type** syms x y **and press Enter.**

   MATLAB creates the two symbolic objects used for this calculation.

2. **Type** f(x, y) = x^2 * sin(y) **and press Enter.**

   This symbolic function accepts two inputs, x and y, and uses them to perform a calculation. Here's the output from this step:

   ```
   f(x, y) =
   x^2*sin(y)
   ```

3. **Type** Result = diff(f) **and press Enter.**

   The output shows the result of the differentiation:

   ```
   Result(x, y) =
   2*x*sin(y)
   ```

   In this case, `Result(x, y)` accepts two inputs, x and y. As before, you can create a picture from the output of `Result()`.

   The example shows the derivative with respect to x, which is the default. To obtain the derivative with respect to y (df/dy), you type `diff(f, y)` instead.

4. **Type** plot(Result(1:50, 1:50)) **and press Enter.**

   Figure 15-3 shows the output of the plot created in this case.

   Notice that in this case, you must provide both x and y inputs, which isn't surprising. However, the two vectors must have the same number of elements or MATLAB will raise an exception.

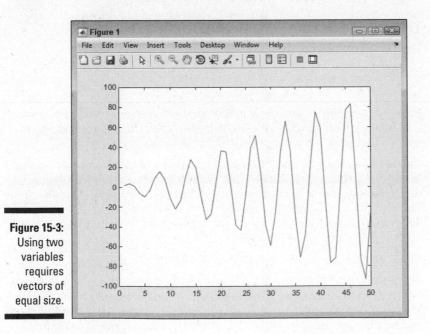

Figure 15-3: Using two variables requires vectors of equal size.

Solving Differential Equations

When working with differential equations, MATLAB provides two different approaches: numerical and symbolic. The following sections demonstrate both approaches to solving differential equations. Note that these sections provide just an overview of the techniques; MATLAB provides a rich set of functions to work with differential equations.

Using the numerical approach

When working with differential equations, you must create a function that defines the differential equation. This function is passed to MATLAB as part of the process of obtaining the result. There are a number of functions you can use to perform this task; each has a different method of creating the output. You can see a list of these functions at http://www.mathworks.com/help/matlab/ordinary-differential-equations.html. The example in this section uses ode23(), but the technique works for the other functions as well.

MATLAB has a specific way of looking at your function. The order in which the variables appear is essential, so you must make sure that your function is created with this need in mind. The example in this section simplifies things to avoid the complexity of many examples online and let you see the process used to perform the calculation. The following steps get you started:

1. **Type** Func = @(T, Y) cos(T*Y) **and press Enter.**

   You see an output of

   ```
   Func =
       @(T,Y)cos(T*Y)
   ```

   Many of the sources you see will tell you that you must place the equation in a separate function file on disk. However, this example demonstrates that creating a temporary function works just fine.

   The requirements for the differential function are that you must provide an input for time and another input containing the values for your equation. The time value, T, is often unused, but you can use it if you want. The variables can consist of anything required to obtain the result you want. In this case, you input a simple numeric value, Y, but inputs can be vectors, matrices, or other objects as well.

2. **Type** [TPrime, YPrime] = ode23(Func, [-10, 10], .2); **and press Enter.**

   When using ode23(), you must provide a function — Func in this case — as input. As an alternative, you provide the name of the file containing the function. The second argument is a vector that contains the starting and ending times of the calculation. The third argument is the starting input value for the calculation.

The TPrime output is always a vector that contains the time periods used for the calculation. The YPrime output is a vector or matrix that contains the output value or values for each time period. In this case, YPrime is a vector because there is only one output value.

3. **Type** plot(TPrime, YPrime) **and press Enter.**

   You see the plotted result for this example, as shown in Figure 15-4.

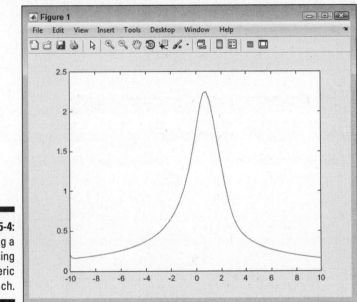

Figure 15-4: Obtaining a result using the numeric approach.

Using the symbolic approach

When working with the symbolic approach, you rely on the functionality of the Symbolic Math Toolbox to speed the solution along and make it a little easier to solve. Even though the solution in the previous section looks easy, it can become quite complicated when you start working with larger problems. The symbolic approach is a little more straightforward. When using the symbolic approach, you rely on dsolve(), which is described at http://www.mathworks.com/help/symbolic/dsolve.html. The following steps show a simple example of using dsolve() to create a differential solution and then plot it:

1. **Type** Solution = dsolve('Dy=(t^2*y)/y', 'y(2)=1', 't') **and press Enter.**

   The arguments to dsolve() consist of the equation you want to solve, the starting point for y (a condition), and the name of the independent variable. You see the following output from this entry:

```
Solution =
t^3/3 - 5/3
```

2. **Type** Values = subs(Solution, 't', -10:.1:10); **and press Enter.**

   `Solution` simply contains the solution to the equation given the conditions you provide. The `subs()` function substitutes values for `t` one at a time. In this case, the values range from –10 to 10 in 0.1 increments. When this command completes, `Values` contains a list of results for the values you provided that you can use as plot points.

3. **Type** plot(Values) **and press Enter.**

   You see the output shown in Figure 15-5.

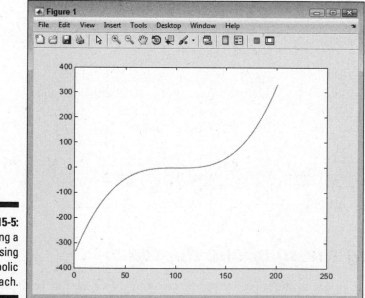

Figure 15-5:
Obtaining a
result using
the symbolic
approach.

Chapter 16

Creating Super Plots

● ●

● ●

*P*lots help people understand data; there is no doubt about that. Previous chapters demonstrate that MATLAB provides excellent plotting features, which only makes sense considering the value of plots to everyone who uses MATLAB. Without plots, trying to explain the math behind a project would be hard — and describing the consequences of the math would be impossible. In order to truly understand math, most people need a picture, and plotting provides that picture.

You may have looked at the plots in the previous chapters and thought that there must be more functionality available. There *is* more — more of everything needed to create just the plot you had in mind! That's what this chapter is about — creating plots that distinguish themselves with regard to clarity and their ability to communicate abstract concepts in clear terms. In other words, this chapter is about plot pizazz.

Of course, misusing the power of plots is easy to do, especially plots that have the features of MATLAB at their disposal. The first section of the chapter helps define precisely what a super plot is, why you'd want to use one, and how to avoid bedazzling your viewers to the point that nothing gets through. In other words, this first section is essential for people who want to create amazing plots without first getting an art degree.

The remainder of the chapter is about actually using advanced plot features. You discover that MATLAB provides access to a number of features that you can use to enhance any given plot. MATLAB also provides specialized plots that help you convey information in new ways — ways other than the omnipresent bar chart beloved by so many businesspeople. (Not that there is anything particularly wrong with a bar chart when you really do need one, but these other plots give you interesting alternatives.) Finally, in this chapter you see how to animate your plots. No, animation is not for entertainment purposes (although, animation could certainly be used to achieve that goal). Rather, animation helps you to view a problem from several perspectives and to understand the solutions needed to address it.

Understanding What Defines a Super Plot

A super plot is actually just a regular plot, but with a lot more pizazz. You add features to a regular plot that emphasize certain elements or make specific concepts clearer. The idea is to use these features to clarify concepts and to help remove the abstractions that cause people to avoid thinking about math most of the time. In short, you use specific features to turn your abstract idea into something concrete — something that others can grasp and understand. Until you get to the point at which others can truly see what you mean with the numbers, trying to convince anyone to see your point of view is useless.

It might be easy at first to think that a super plot is meant to bedazzle the audience — to hide bad numbers or simply impress people in a manner that makes them less likely to disagree. However, in this day of dramatic special effects that only slightly faze people, a plot with amazing graphics isn't going to achieve much. So, a super plot isn't meant to hide anything.

A problem for most people embarking on a journey of intense graphic manipulation for the first time is that the eye candy does become quite addictive. You end up wanting to add just about everything to your plot. The result is often hideous and sometimes comical, but it isn't convincing. Here are some things you can do to keep your plots from becoming a little too intense:

- ✔ Avoid using too many plot extras on a single plot. Use only those plot extras that truly enhance the numbers in a way that makes them easier to understand.

- ✔ Create a consistent look across plots so that you don't end up with plots that are more distracting than informative.

- ✔ Rely on labels and other graphic elements (such as text boxes) to help explain the data whenever necessary.

- ✔ Use the same *rotation* (the same perspective or point of view) in associated plots whenever possible. (You see how to use rotation in Chapter 7, and several other chapters work with rotation in various ways. Later in this chapter you also see rotation-type effects.)

✔ Create specialized plots only when the specialized version conveys information in a manner that a standard plot doesn't. (People are often distracted by the special plots.)

✔ Employ animation carefully to show a numeric sequence or to allow viewing of a problem from all sides. Avoid using animation as a special effect.

✔ Define the target of emphasis for each plot before you actually add any emphasis so that you can focus on just the area that needs adjustment.

✔ Develop a strategy for presenting the information before you actually create the plots so that you have a clear plan in mind.

Using the Plot Extras

Chapters 6 and 7 of the book focus attention on the basics of creating and using plots to convey information to others. You see a few extras used in the presentation of data in Chapter 12. The following sections go beyond the materials in these other chapters to describe the kinds of extras you can use for emphasizing specific data. For example, the way in which you configure the grid can help make differences between data more obvious.

You can always clear the current figure from the window by typing **cla** and pressing Enter. This command clears all the plot information from the figure, but doesn't close the figure window.

Using grid()

The grid helps you see how data interrelates on a plot. However, sometimes the grid isn't in the right place; it's too large, or simply in the way. The following grid() functions help you use a grid effectively on your plot:

✔ grid('on'): Turns the grid on so that it appears in the current figure.

✔ grid('off'): Turns the grid off so that you can't see it in the current figure.

✔ grid(): Toggles the grid between on and off.

✔ grid('minor'): Toggles the minor grid ticks on and off.

Obtaining the current axis using gca

Many of the functions used to enhance the appearance of a plot require access to the current axis. However, when you create a plot, the only handle you have is the one to the plot itself. To obtain the axis handle, you type something like `Bar1Axis = gca` and press Enter. The result is that you now have an axis handle in `Bar1Axis`.

Creating axis dates using datetick()

You use `datetick()` to add dates to a plot axis. When using `datetick()`, you need an axis that has numbers that are in the range of the dates you need. For example, when you type **datenum('9,15,2014')** and press Enter, you get an output value of `735857`. When `datetick()` sees this value, it converts the number to a date.

The `datenum()` function also accepts time as input. When you type **datenum ('09/15/2014 08:00:00 AM')** and press Enter, you get `7.3586e+05` as output. Notice that the integer portion of the value is the same as before, but the decimal portion has changed to show the time. If you don't provide a time, the output is for midnight of the day you select. You can convert a numeric date back to a string date using the `datestr()` function.

The x-axis in this example uses date values. To create an x-axis data source, type **XSource = linspace(datenum('09/15/2014'), datenum('09/19/2014'), 5);** and press Enter. This act creates a vector that contains the dates from 09/15/2014 to 09/19/2014. The `linspace()` function returns a vector that contains the specified number of value (5 in this case) between the two values you specify.

To create the y-axis data source, type **YSource = [1, 5, 9, 4, 3];** and press Enter. Type **Bar1 = bar(XSource, YSource)** and press Enter to create the required plot. The default tick spacing will show too many points, so type **set(gca, 'XTick', lins pace(datenum('09/15/2014'), datenum('09/19/2014'), 5))** and press Enter to set the tick spacing. Figure 16-1 shows how your plot should look. Notice that the x-axis doesn't use the normal numbering scheme that begins with 1 — it uses a date number instead (expressed as an exponent rather than an integer). Even though the x-axis numbers look the same, you see in the next paragraph that they aren't.

To turn the x-axis labels into dates, you now use the `datetick()` function. Type **datetick('x', 'dd mmm yy', 'keeplimits', 'keepticks')** and press Enter. Figure 16-2 shows the plot with dates in place.

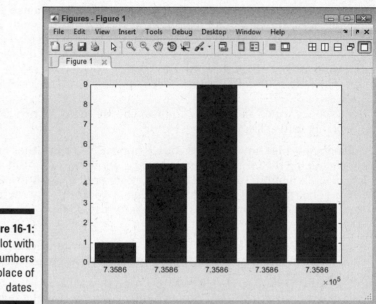

Figure 16-1:
A plot with numbers in place of dates.

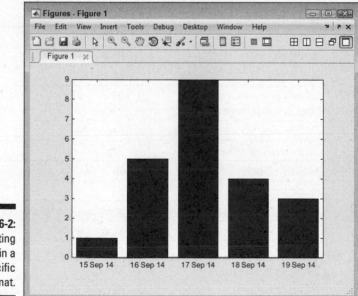

Figure 16-2:
Creating dates in a specific format.

All the arguments used with datetick() are optional. When you use datetick() by itself, the output appears on the x-axis using a two-digit month and a two-digit day. The end points also have dates, so instead of seeing just five dates, you see seven (one each for the ends). The example uses the following arguments in this order to modify how datetick() normally works:

- ✔ **Axis:** Determines which axis to use. You can choose the x-, y-, or z-axis (when working with a 3D plot).

- ✔ **Date format:** Specifies how the date should appear. You can either use a string containing the format as characters, as shown in Table 16-1, or you can use a number to choose a default date option, as shown in Table 16-2. (The two tables assume a datenum() value of '09/15/2014 08:00:00 AM'.) The example uses a custom format, so one of the numeric options won't work.

- ✔ **'keeplimits':** Prevents MATLAB from adding entries to either end of the axis. This means that the example plot retains five x-axis entries rather than getting seven.

- ✔ **'keepticks':** Prevents MATLAB from changing the value of the ticks.

Table 16-1   Strings Used to Create a Date Format

String	Purpose	Example
yyyy	Four-digit year	2014
yy	Two-digit year	14
QQ	Quarter of the year using the letter Q and a single number	Q1
mmmm	Month using full name	September
mmm	Three-letter month name	Sep
mm	Two-digit month	09
m	Single-letter month name	S
dddd	Day using full name	Monday
ddd	Three-letter day name	Mon
dd	Two-digit day	15
d	Single-letter day name	M
HH	Two-digit hour	08 when no AM/PM used, 8 AM otherwise
MM	Two-digit minutes	00

String	Purpose	Example
SS	Two-digit seconds	00
FFF	Three-digit milliseconds	000
AM or PM	AM or PM is used rather than 24-hour military time	8:00:00 AM

Table 16-2   Numeric Selections for Standardized Dates

Number	String Format Equivalent	Example
-1 (default)	`'dd-mmm-yyyy HH:MM:SS'` or `'dd-mmm-yyyy'` (no time output at midnight)	15-Sep-2014 08:00:00 or 15-Sep-2014
0	`'dd-mmm-yyyy HH:MM:SS'`	15-Sep-2014 08:00:00
1	`'dd-mmm-yyyy'`	15-Sep-2014
2	`'mm/dd/yy'`	09/15/14
3	`'mmm'`	Sep
4	`'m'`	S
5	`'mm'`	09
6	`'mm/dd'`	09/15
7	`'dd'`	15
8	`'ddd'`	Mon
9	`'d'`	M
10	`'yyyy'`	2014
11	`'yy'`	14
12	`'mmmyy'`	Sep14
13	`'HH:MM:SS'`	08:00:00
14	`'HH:MM:SS PM'`	8:00:00 PM
15	`'HH:MM'`	08:00
16	`'HH:MM PM'`	8:00 PM
17	`'QQ-YY'`	Q3-14
18	`'QQ'`	Q3
19	`'dd/mm'`	15/09

(continued)

Table 16-2 *(continued)*

Number	String Format Equivalent	Example
20	`'dd/mm/yy'`	15/09/14
21	`'mmm.dd,yyyy HH:MM:SS'`	Sep.15,2014 08:00:00
22	`'mmm.dd,yyyy'`	Sep.15,2014
23	`'mm/dd/yyyy'`	09/15/2014
24	`'dd/mm/yyyy'`	15/09/2014
25	`'yy/mm/dd'`	14/09/15
26	`'yyyy/mm/dd'`	2014/09/15
27	`'QQ-YYYY'`	Q3-2014
28	`'mmmyyyy'`	Sep2014
29	`'yyyy-mm-dd'` (ISO 8601)	2014-09-15
30	`'yyyymmddTHHMMSS'` (ISO 8601)	20140915T080000
31	`'yyyy-mm-dd HH:MM:SS'`	2014-09-15 08:00:00

Creating plots with colorbar ( )

Using a color bar with your plot can help people see data values based on color rather than pure numeric value. The color bar itself can assign human-understandable values to the numeric data so that the data means something to those viewing it. The best way to work with color bars is to see them in action. The following steps help you create a color bar by using the `colorbar()` function and use it to define values in a bar chart:

1. **Type** YSource = [4, 2, 5, 6; 1, 2, 4, 3]; **and press Enter.**

   MATLAB creates a new data source for the plot.

2. **Type** Bar1 = bar3(YSource); **and press Enter.**

   You see a new bar chart like the one shown in Figure 16-3. Even though the data is in graphic format, it's still pretty boring. To make the bar chart easier to work with, the next step changes the y-axis labels.

3. **Type** CB1 = colorbar('EastOutside'); **and press Enter.**

You see a color bar appear on the right side of the plot, as shown in Figure 16-4. You can choose other places for the color bar, including inside the plot. Don't worry about the color bar ticks not matching those of the bar chart for now.

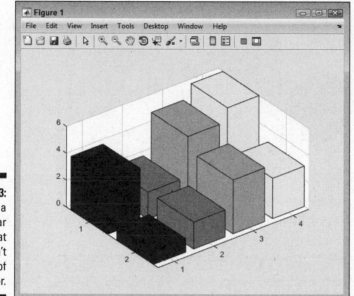

Figure 16-3:
Defining a basic bar chart that doesn't make use of color.

4. **Type the following code into the Command window, pressing Enter after each line.**

```
for Element = 1:length(Bar1)
    ZData = get(Bar1(Element),'ZData');
    set(Bar1(Element), 'CData', ZData,...
        'FaceColor', 'interp')
end
```

A number of changes take place. The bars are now colored according to their value. In addition, the ticks on the color bar now match those of the bar chart as shown in Figure 16-5. However, the color bar just contains numbers, so it doesn't do anything more than the y-axis labels do to tell what the colors mean.

5. **Type** set(CB1, 'YTickLabel', {'', 'Awful', 'OK', 'Better', 'Average', 'Great!', 'BEST'}); **and press Enter.**

The chart now has meanings assigned to each color level, as shown in Figure 16-6.

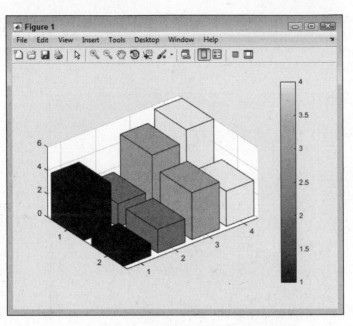

Figure 16-4:
The color bar appears on the right side of the plot.

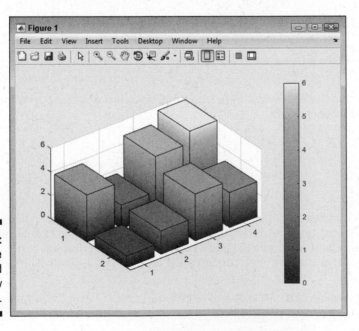

Figure 16-5:
The bars are now colored to show their values.

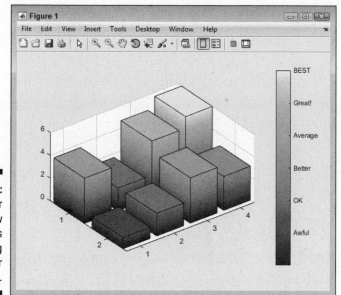

Figure 16-6:
The color
bar now
conveys
meaning
to the bar
chart.

The color scheme that MATLAB uses by default isn't the only color scheme available. The `colormap()` function lets you change the colors. For example, if you type **colormap('cool')** and press Enter, the colors change appropriately. You can also create custom color maps using a variety of techniques. To get more information, see the `colormap()` documentation at `http://www.mathworks.com/help/matlab/ref/colormap.html`.

Interacting with daspect

How the 3D effect appears onscreen depends on the data aspect ratio. The `daspect()` function lets you obtain the current aspect ratio and set a new one. The aspect ratio is a measure of how the x-, y-, and z-axis interact. For example, an aspect ratio of [1, 2, 3] would mean that for every 1 unit of the x-axis, there are two units of the y-axis and three units of the z-axis. Perform the following steps to see how this feature works:

1. Type YSource = [1, 3, 5; 3, 7, 9; 5, 7, 11]; **and press Enter.**

MATLAB creates a data source for you.

2. Type Bar1 = bar3(YSource); **and press Enter.**

You see a 3D bar chart appear.

3. Type rotate(Bar1, [0, 0, 1], 270); **and press Enter.**

The bar chart rotates so that you can see the individual bars easier, as shown in Figure 16-7.

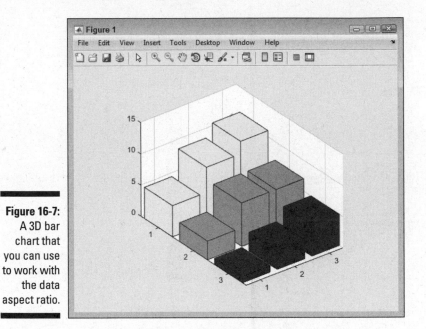

Figure 16-7:
A 3D bar
chart that
you can use
to work with
the data
aspect ratio.

4. **Type** daspect() **and press Enter.**

   The output contains three values, like this:

   ```
   ans =
        0.3571     0.2679     2.1670
   ```

   So, you now know the current aspect ratio of the plot, with the first
   number representing the x-axis value, the second number the y-axis
   value, and the third number the z-axis value. Your numbers may not precisely match those shown in the book.

5. **Type** daspect([.25, 1, 1.2]); **and press Enter.**

   The data aspect ratio changes to create tall, skinny-looking bars like those
   shown in Figure 16-8. Compare Figures 16-7 and 16-8 and you see that the
   differences between the individual bars appears greater, even though
   nothing has changed. The data is precisely the same as before, as is the
   rotation, but the interpretation of the data changes.

6. **Type** daspect([.65, .5, 7]); **and press Enter.**

   The impression is now that the differences between the data points are
   actually quite small, as shown in Figure 16-9. Again, nothing has changed
   in the data or the rotation. The only thing that has changed is how the
   data is presented.

7. **Type** daspect('auto') **and press Enter.**

   The data aspect returns to its original state.

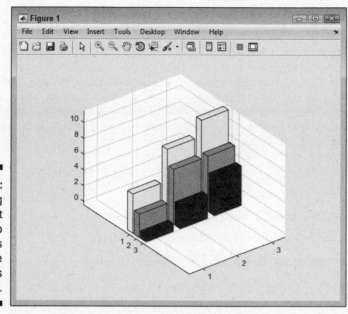

Figure 16-8:
Modifying the aspect ratio changes how the data is perceived.

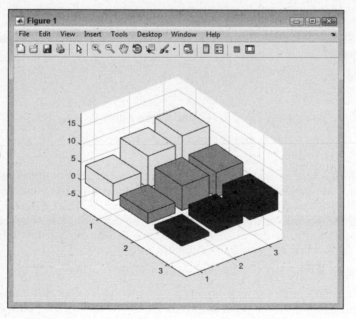

Figure 16-9:
You can make the differences between data points appear great or small.

Interacting with pbaspect

The previous section tells how to modify the data aspect ratio. In addition, you see a number of examples that show how to use rotation to modify the appearance of the data. This section discusses the plot box aspect ratio. Instead of modifying the data, the plot box aspect ratio modifies the plot box — the element that holds the plot in its entirety — as a whole. The appearance of the data still changes, but in a different way than before. The following steps get you started with this example:

1. **Type** YSource = [1, 3, 5; 3, 7, 9; 5, 7, 11]; **and press Enter.**

   MATLAB creates a data source for you.

2. **Type** Bar1 = bar3(YSource); **and press Enter.**

   You see a 3D bar chart appear.

3. **Type** rotate(Bar1, [0, 0, 1], 270); **and press Enter.**

   The bar chart rotates so that you can see the individual bars easier. (Refer to Figure 16-7.)

4. **Type** pbaspect() **and press Enter.**

   As before, you get three values: x-, y-, and z-axis. However, the numbers differ from before because now you're working with the plot box aspect ratio and not the data aspect ratio. Here are typical output values at this point:

   ```
   ans =
        2.8000     4.0000     2.4721
   ```

5. **Type** pbaspect([1.5, 1.5, 7]); **and press Enter.**

   The differences between the data points seem immense, as shown in Figure 16-10.

   Notice how changing the plot box aspect ratio affects both the plot box and the data so that the plot box no longer is able to change settings, such as the spacing between bars (as shown in Figures 16-8 and 16-9). What this means in particular is that you don't have to worry about bars ending up outside the plot area and not being displayed. The bars and the plot box are now locked together.

6. **Type** pbaspect([4, 5, 1]); **and press Enter.**

   The data points now seem closer together, even though nothing has changed in the data, as shown in Figure 16-11. At this point, it helps to compare Figures 16-7 through 16-11. These figures give you a better idea of how aspect ratio affects the perception of your data in various ways.

7. Type pbaspect('auto'); **and press Enter.**

The plot aspect returns to its original state.

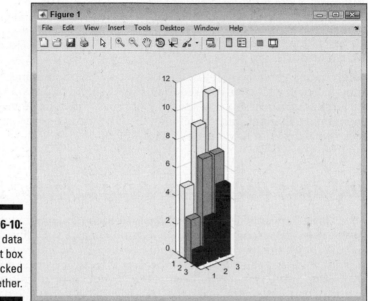

Figure 16-10:
The data
and plot box
are locked
together.

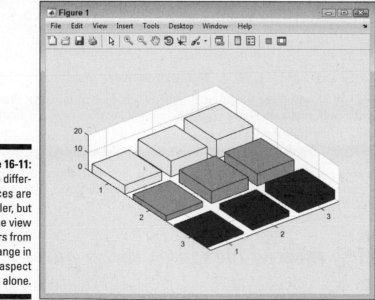

Figure 16-11:
The differ-
ences are
smaller, but
the view
differs from
a change in
data aspect
alone.

Working with Plot Routines

Previous chapters discuss some of the more common plots. These are the sorts of plots you might find anywhere, such as a bar chart. Just about everyone uses them, so they're good plots to choose for your data as well. Familiarity with the form lets people focus on the data rather than the plot. Use one of these common plots whenever possible.

Unfortunately, using a common plot is not always possible. Sometimes you need to model data in a manner that requires a special kind of plot. MATLAB provides a large assortment of these special plot types. The following sections demonstrate a number (but not all) of them. If you don't see what you want, check Appendix B for a complete listing of plots.

Finding data deviations using errorbar ( )

An error bar chart shows the level of confidence in each data point along a line. What you see is the transition between data points as a solid line, and then a bar that shows the amount of deviation, both positive and negative, from that point. A viewer can tell whether the data is truly accurate or simply a guess by looking at the amount of deviation for particular points along the line. This kind of plot helps with analysis in determining whether the data is solid enough to rely on or whether additional input is needed to obtain a better estimate. The following steps help you create an error bar plot:

1. **Type** YSource = [1, 2, 4, 7, 5, 3]; **and press Enter.**

   These values represent the actual data points. They tell the viewer what you think the actual values might be.

2. **Type** ESource = [.5, 1, 1, 1.5, 1, .5]; **and press Enter.**

   These values represent the amount of error for each data point. This is the plus or minus amount and tells the viewer how confident you are about the values in YSource.

3. **Type** EBar1 = errorbar(YSource, ESource); **and press Enter.**

   This step creates the error bar plot shown in Figure 16-12. Notice that each data point has an error bar associated with it, showing the potential amount of deviation for that data point.

Ranking related measures using pareto ( )

The Pareto diagram was originally created by Vilfredo Pareto in early 1900s Europe. It shows the relations of measures in decreasing order of occurrence. By using a Pareto diagram, it becomes possible to see where to spend most of your time looking for a problem. You can read more about Pareto diagrams at http://www.pqsystems.com/qualityadvisor/DataAnalysisTools/ pareto_diagram.php.

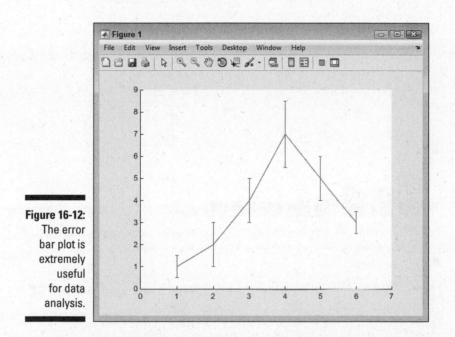

Figure 16-12:
The error bar plot is extremely useful for data analysis.

To create this example, you begin by defining a data source with multiple occurrences of some values. Type **YSource = [1, 2, 4, 2, 6, 2, 3, 4, 1, 2];** and press Enter to define the data source. Now that you have a data source, create the Pareto diagram by typing **Par1 = pareto(YSource);** and pressing Enter. Figure 16-13 shows typical output from this example.

Plotting digital data using stairs ()

The stairstep plot is commonly used to display time history plots of digitally sampled systems. The plot shows the result of continuous change, but also defines the interval at which these changes occur as a precise time order rather than a continuous change as you see in analog systems. The following steps help you create a stairstep plot:

1. **Type** YSource = [1, 2, 4, 7, 5, 3]; **and press Enter.**

   These values represent the actual data points. They tell the viewer what you think the actual values might be.

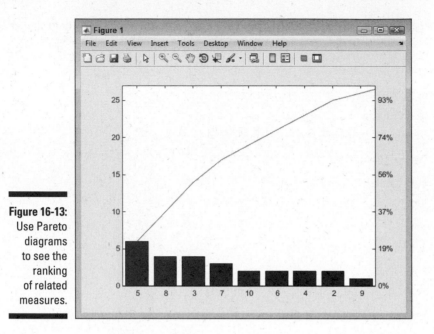

2. **Type** XSource = datenum('9/15/2014'):1:datenum('9/20/2014'); **and press Enter.**

   These values represent dates on which the data points were collected. In this case, the range is from 9/15/2014 to 9/20/2014 with a single day between values. The output is a vector containing six date values.

3. **Type** Stair1 = stairs(XSource, YSource); **and press Enter.**

   You see the start of the plot. However, the dates are still numbers, and MATLAB has provided a half-day interval instead of the one-day interval you might like.

4. **Type** set(gca, 'XTick', XSource); **and press Enter.**

   The dates now appear as a single date for each data point, and there is a one-day interval between each date.

5. **Type** datetick('x', 6, 'keeplimits', 'keepticks'); **and press Enter.**

   The plot now contains actual date strings that contain the month and day. This is the number 6 selection from Table 16-2. The completed plot appears in Figure 16-14.

Showing data distribution using stem ( )

A stem plot, which shouldn't be confused with a stem and leaf display (see http://www.purplemath.com/modules/stemleaf.htm for details), shows a distribution of numbers across a range using the x-axis as the starting point. Each data point appears as the endpoint of a line that goes from the x-axis to the value of that particular data point. The stem plot has a number of uses, especially when used in the 3D form. For example, you could use it to track particle motion (as described at `http://people.rit.edu/pnveme/pigf/ThreeDGraphics/thrd_bp_stem.html`).

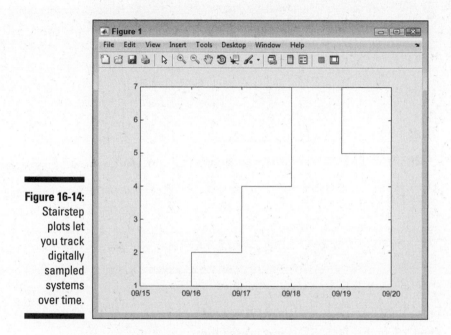

Figure 16-14:
Stairstep plots let you track digitally sampled systems over time.

To create a stem plot, begin by defining a data source. Type **YSource = [-10:1:10];** and press Enter. Type **stem(YSource);** and press Enter to create the actual plot. However, you will find that the y-axis limits usually aren't set to see all the endpoints properly. Fix this issue by typing **set(gca, 'YLim', [-11, 11]);** and pressing Enter. Figure 16-15 shows typical output for this plot.

Drawing images using fill

You can create images using MATLAB. All you need is a mathematical model that describes the points used to describe the image shape. For example, to draw a square, you simply provide the x- and y-axis coordinate for each corner. You can see a number of these shapes demonstrated at

http://www.mathworks.com/matlabcentral/fileexchange/35293-matlab-plot-gallery-fill-plot/content/html/Fill_Plot.html.
The following steps help you create an image of your own.

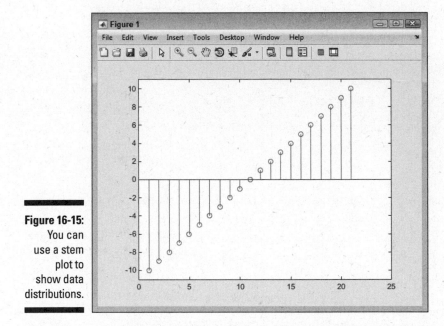

Figure 16-15:
You can use a stem plot to show data distributions.

1. **Type** XSource = [1, 1, 5, 5]; **and press Enter.**

2. **Type** YSource = [1, 5, 5, 1]; **and press Enter.**

   The XSource and YSource variables contain coordinates to draw a square. The lower-left corner is at 1,1; the upper-left corner is at 1,5; the upper-right corner is at 5,5; and the lower-right corner is at 5,1.

3. **Type** fill(XSource, YSource, 'b'); **and press Enter.**

   MATLAB creates the image, but the image consumes the entire drawing area. Notice that the image is filled with blue. You can choose any color you like using the options in Table 16-3. To see the image set apart from the plot area, you need to change the x and y limits.

4. **Type** set(gca, 'XLim', [0, 6]); **and press Enter.**

5. **Type** set(gca, 'YLim', [0, 6]); **and press Enter.**

   The image is now clear and in the center of the plot, as shown in Figure 16-16.

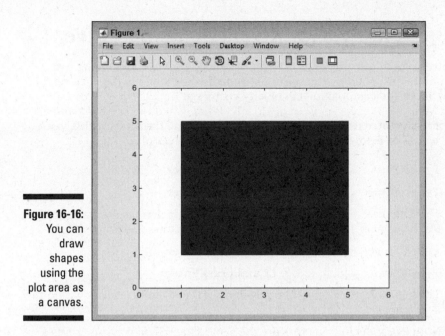

Figure 16-16:
You can draw shapes using the plot area as a canvas.

Selecting a color for your plot is important. Table 16-3 contains a listing of the most common colors — those you can specify using a color letter or descriptive name. However, you can specify partial RGB values. For example, an RGB value of [.5, .25, 0] provides a nice brown. Each entry for red, green, and blue must have a value between 0 and 1.

Table 16-3	Color Choices for fill()	
RGB Value	*Color Letter*	*Description*
[1 1 0]	y	yellow
[1 0 1]	m	magenta
[0 1 1]	c	cyan
[1 0 0]	r	red
[0 1 0]	g	green
[0 0 1]	b	blue
[1 1 1]	w	white
[0 0 0]	k	black

Displaying velocity vectors using quiver ( )

A quiver plot shows the velocity vectors defined by the components u and v at points described by the coordinates defined by x and y. (You can read more about the composition of velocity vectors at http://mathworld. wolfram.com/VelocityVector.html.) When you don't specify x and y, the plot is created using equally spaced values along the x-axis with a value of y = 1. The following steps help you create a quiver plot.

1. **Type** XSource = [1, 1, 1, 1, 1, 1]; **and press Enter.**

2. **Type** YSource = [1, 1, 1, 1, 1, 1]; **and press Enter.**

   These two vectors define the x/y coordinate pairs used as the starting point for the vectors. In this case, all the vectors have the same origin of 1, 1.

3. **Type** USource = [1, 2, 3, 4, 5, 6]; **and press Enter.**

4. **Type** VSource = [6, 5, 4, 3, 2, 1]; **and press Enter.**

   These two vectors define the positions of the velocity vectors within the x-y plane.

   Notice that the four vectors are the same size. Whenever you create a quiver plot, the vectors must be the same size because they act in pairs to create coordinates.

5. **Type** quiver(XSource, YSource, USource, VSource); **and press Enter.**

   You see a quiver plot similar to the one shown in Figure 16-17.

Displaying velocity vectors using feather ( )

A feather plot is similar to the quiver plot, except that it draws the vectors evenly spaced along the x-axis. To see this plot in action, type **USource = [-6:1:6];** and press Enter to create u; type **VSource = [6:-1:-6];** to create v. Type **feather(USource, VSource)** and press Enter to see a plot similar to the one shown in Figure 16-18.

Displaying velocity vectors using compass ( )

The concept behind a compass plot is that it provides a similar view to a feather and quiver plot, but all the values emanate from a single starting point and the output is a kind of polar plot. The values of u and v create Cartesian coordinates. (See the description of Cartesian coordinates at http://www.

`mathsisfun.com/data/cartesian-coordinates.html` for more details.)
A compass plot could be used to display directional data, such as wind direction and velocity. (See the discussion at `http://dali.feld.cvut.cz/ucebna/matlab/techdoc/umg/chspec20.html` for more details.)

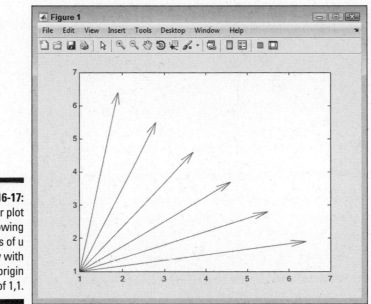

Figure 16-17:
A quiver plot showing values of u and v with an origin of 1,1.

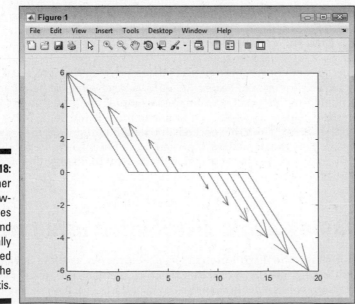

Figure 16-18:
A feather plot showing values of u and v equally spaced along the x-axis.

To see this plot in action, type **USource = [1, -4, 3, -7, 8, -9, 2, 4, -2, 3, -5, 8, 9];** and press Enter to create u; type **VSource = [12:-2:-12];** to create v. Type **compass(USource, VSource)** and press Enter to see a plot similar to the one shown in Figure 16-19.

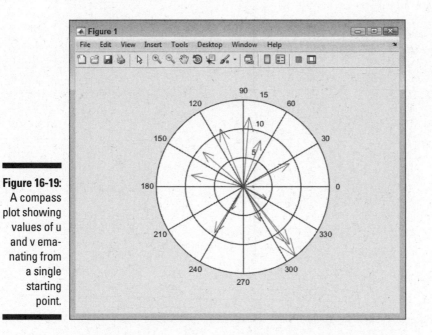

Figure 16-19: A compass plot showing values of u and v emanating from a single starting point.

Working with polar coordinates using polar ()

A polar plot accepts polar coordinates (see http://www.mathsisfun.com/polar-cartesian-coordinates.html for a discussion of the difference between polar and Cartesian coordinates) as input, and plots them in the Cartesian plain using a polar plot. The value of theta is the angle between the x-axis and the vector. The value of rho is the length of the vector. To see this plot in action, type **theta = 0:0.01:2*pi;** and press Enter to create theta; type **rho = 1 - theta;** to create rho. Type **polar(theta, rho)** to display the plot shown in Figure 16-20.

Displaying angle distribution using rose ()

A rose plot shows the distribution of angles expressed as theta in radians. The output is an angle histogram that relies on a polar plot, where the angle of each histogram is controlled by the angle of theta. You can define the number of bins used to compute the distribution. The default number of bins is 20. To see this plot in action, type **theta = 0:0.01:2*pi;** and press Enter to create theta. Type **rose(theta, 36)** to display the plot shown in Figure 16-21.

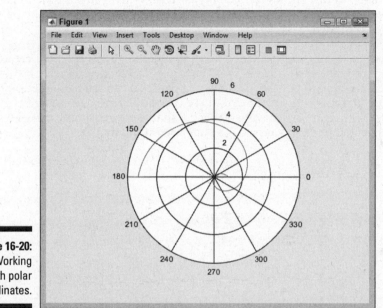

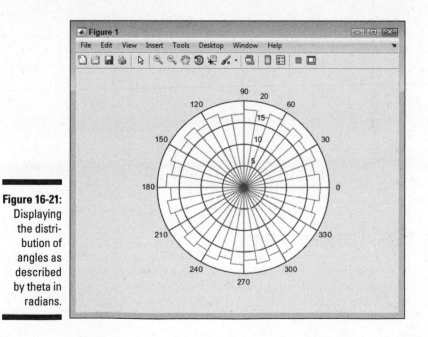

Spotting sparcity patterns using spy()

The spy plot accepts a sparse matrix as input, analyzes it for a pattern, and displays a plot that shows the sparcity pattern. When you supply a full matrix (one that doesn't contain empty elements), the spy plot analyzes the pattern of nonzero elements.

The example in this section uses a mystery function, `bucky()`, that creates a sparse matrix for `spy()` to analyze. The `bucky()` function actually creates a sparse matrix based in what amounts to a ball. You can read about `bucky()` at `http://matlab.izmiran.ru/help/techdoc/math/sparse12.html`. Relying on `bucky()` whenever you need a sparse matrix simply makes testing easier because `bucky()` provides a standard output matrix.

To see this example work, type **spy(bucky());** and press Enter. You see the output shown in Figure 16-22.

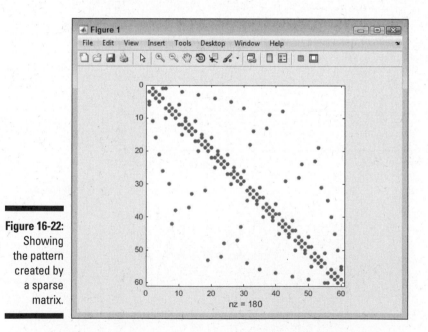

Figure 16-22: Showing the pattern created by a sparse matrix.

Employing Animation

Animation presents a story of data changes over time. When you look at a standard plot, what you see is a snapshot of data at a specific time — the time the plot was created. It could be that the data won't change any time soon or possibly at all. However, most data does change. When it does, you need to decide whether presenting a series of views of these changes is in the viewer's interest. If so, then animation becomes part of your display strategy.

It's easy to misuse animation or to use it poorly. An example of a misuse of animation is when the animation becomes the focus of the presentation, rather than the data. Displaying the single set of test results from a recent experiment with flashing bars is an example of animation misuse because the flashing bars, rather than the data, become the presentation. Poor animation use is when you choose a method of presentation that detracts from the data even though the data is the focus of the presentation. For example, when the screen updates too slowly, the animation, not the data, has become the focus of the presentation.

MATLAB provides access to three kinds of animation. Choosing the right kind of animation for your particular need is important. The following list describes the kinds of animation and when you typically use them:

- ✔ **Static image playback:** The oldest and most common form of animation is the playback of a series of static images. When you go to the theater, you really do see a series of static images — one after another — presented in rapid succession, creating the illusion of a continuous stream of information. The movie approach is best used for any sort of complex plot presentation. The screen can update quickly, creating smooth animation. The downside to this approach is that it isn't dynamic. You still rely on static data, which means that you can't provide updates during the presentation.

- ✔ **Object updates:** Modifying the object properties is another way to create animation. You can adjust data values, color schemes, and other object properties to create the illusion of animation. Of the three approaches, this is the most flexible and the least reliant on outside resources. In addition, it's quite fast. The downside is that you need to write a lot of code to make this technique work. In addition, the information tends to be more static than the third approach, but less static than the first. You can create data updates, but doing so requires writing more code, which is time-consuming and inconvenient at times (as contrasted to the automation provided by the other methods).

- ✔ **Data updates:** It's possible to connect plot objects directly to data sources. As the data source changes, so does the plot. Of the three approaches, this one is the most dynamic. You commonly use this approach to model real-time data (information that is actually changing during your presentation). However, you don't have as much control over the presentation of the data as you do with the second approach. The downside to this approach is that you normally need an external data source to make it effective — that is, to truly get the full benefit of using this approach. Unfortunately, presentations often suffer from glitches such as loss of connectivity.

Working with movies

The static image playback approach, also called a movie, requires that you grab a series of screenshots of your data as it changes by calling `getframe()`. Most of the examples of using `getframe()` show it being used to grab the default object, which are the axes. However, you can supply a handle to any object and make it the focus of your movie. In addition, you can specify that only part of the target object appear in the movie by specifying a cropping rectangle using one of the arguments. This feature lets you do things like create wipe effects, where you focus in on one object and progressively reveal the rest of the plot from there by changing the rectangular settings.

The frames are placed in a matrix. After you have collected enough frames, you can play your movie using the `movie()` function. This function accepts a number of inputs, but the three most common are the matrix holding the data to play, the number of times you want to play the movie, and the rate at which to play the movie (the frames per section, or fps). Listing 16-1 shows a typical use of the `getframe()` and `movie()` functions. You can also find this script in the `MakeMovie.m` file supplied with the downloadable source code.

Listing 16-1:   Creating a Movie Animation

```
YSource = [1, 2, 5; 2, 4, 8; 7, 9, 10];
Bar1 = bar3(YSource);
rotate(Bar1, [0, 0, 1], 270)

FigHandle = gcf();

for Frame = 1:32
    Frames(Frame) = getframe(FigHandle,...
        [0, 0, 15 * Frame, 15 * Frame]);
end

clf
movie(FigHandle, Frames, 1, 5);
```

The code begins by creating a 3D bar chart and rotating it so that you can easily see the bars. The `rotate()` function accepts three arguments in this case: the handle of the bar chart; a vector containing indicators of which axis to turn (x, y, and z); and the amount to rotate the bar chart in degrees. In this case, the plot rotates around the z-axis.

You can add a call to `rotate3d('on')` to allow mouse-based rotation of the figure by the user. When you no longer want to allow rotation, call `rotate3d('off')` instead. Calling `rotate3d()` by itself toggles between the on and off state. When you supply a handle to the `rotate3d()` function, the changes affect the figure pointed to by the handle rather than the current figure.

The next step is to generate the movie data. The data consists of 32 frames of data. Each loop obtains information from the figure as a whole, starting in the lower-left corner of the figure. The width and height of the screenshot increases with each loop, so each screenshot is a little bigger and shows a little more of the image as a whole.

After creating the movie, the code clears the screen and then calls `movie()` to display the movie onscreen. The screenshots are grabbed using the figure handle, not the axis handle, so the movie must also be played using the figure handle. This is the first argument to `movie()`. The next argument is the movie matrix itself. The final two arguments determine the number of times to play the movie (once) and the frame rate to use (5 fps).

Working with objects

You have all sorts of ways to interact with plot data directly. Previous examples in the book demonstrate just a few of them. By using the `pause()` function, you can create an animation of these changes. Listing 16-2 shows such an example. In this case, the bars in the first row of the 3D bar chart change color one at a time. You could use such an effect during a presentation to bring focus to a particular data item. You can also find this script in the `ChangeObjects.m` file supplied with the downloadable source code.

Listing 16-2: Creating Animation Using Object Changes

```
YSource = [1, 2, 5; 2, 4, 8; 7, 9, 10];
Bar1 = bar3(YSource);
rotate(Bar1, [0, 0, 1], 270)

Colors = get(Bar1(1), 'CData');

for i = 1:6:18
    Colors(i,:) =[2, 2, 2, 2];
    Colors(i+1,:) =[2, 2, 2, 2];
    Colors(i+2,:) =[2, 2, 2, 2];
    set(Bar1(1), 'CData', Colors);
    pause(2);

    Colors(i,:) =[1, 1, 1, 1];
    Colors(i+1,:) =[1, 1, 1, 1];
    Colors(i+2,:) =[1, 1, 1, 1];
    set(Bar1(1), 'CData', Colors);
    pause(2);
end
```

The code begins by creating a 3D bar chart and rotating it so that you can see the data clearly. The code then obtains the colors used to fill the various bars from the CData property and places these values in Colors.

For this example, the CData property is an 18 × 4 matrix containing the colors for each face of the bar chart. Each bar in a chart uses six rows in the CData property. So if this chapter had four bars instead of three, the CData property would have 24 rows rather than 18. The colors for the first bar are contained in rows 1, 2, and 3. The second bar colors appear in 7, 8, and 9, while the third bar colors appear in rows 13, 14, and 15. The other three rows in each set are hidden from view. There is a separate CData property for each of the bar groups. The example works only with the first group, Bar1(1).

Notice how this example uses the for loop. The values show that there are 18 elements. Each bar consumes six elements, so this for loop skips six elements. Instead of having the for loop set i to 1, 2, and 3, this for loop sets the values to 1, 7, and 13, which is precisely what is needed to set the color values in Colors. After it changes the color in each of the three face elements for a particular row, the code goes on to the next row by adding 1 or 2 to the value of i. The final step is to then set the CData value in Bar1(1). You see the change occur onscreen: The bar is highlighted. After a two-second pause, the colors are returned to their original state.

Performing data updates

Directly changing a data source is another way to create animation. The technique involves creating a link between the plot data and the data source. You can create a local data source, such as a variable, to create the animation, but this technique is more likely used with external data sources. Listing 16-3 shows how you can create animation using this approach. You can also find this script in the ChangeData.m file supplied with the downloadable source code.

Listing 16-3:   Creating Animation Using Data Changes

```
YSource = [2, 0, 1, 4, 5, 2, 3];
Bar1 = bar(YSource);

set(Bar1, 'YDataSource', 'YSource');
set(gca, 'YLim', [0, 8]);

for i = 2:7
    YSource(3) = i;
    pause(2);
    refreshdata;
end
```

The code begins with a simple bar chart. It then assigns the bar chart's `YDataSource` to a variable, `YSource`. However, a change to `YSource` doesn't automatically show up on the plot; you must also call `refreshdata`, as shown later in the code.

Adjusting the display of a plot is often necessary to accommodate the animation you provide. In this example, the plot would need to adjust for the higher data values added during the animation. The effect is disruptive because the viewer would focus on these adjustments rather than the data. The call to change the `YLim` value eliminates this problem. Make certain that you check for adjustments of this sort in your own code.

The `for` loop modifies one of the values in `YSource`, waits for two seconds, and then calls `refreshdata` to make the change appear in the plot. What you see onscreen is an increase in one of the bars as the values change. The change simulates data changes that you might see in a real-world, real-time display.

Part VI
The Part of Tens

Enjoy an additional Part of Tens article about ten amazing MATLAB add-ons at
`http://www.dummies.com/extras/matlab`.

In this part . . .

- ✔ See how other people use MATLAB to perform math-related tasks.

- ✔ Discover how others use MATLAB in areas you never thought possible.

- ✔ Get a list of places you could work with your new found MATLAB skills.

- ✔ Consider the job titles associated with having MATLAB knowledge.

Chapter 17

Top Ten Uses of MATLAB

MATLAB is used in a lot of different ways by lots of people in occupations you might not necessarily think about when it comes to a math product. In fact, math is used in ways that many people don't consider. For example, video games simply wouldn't exist without a lot of relatively complex math. However, even the chair you're sitting in required some use of math to ensure that it would perform as expected. A mission to Mars or to the bottom of the ocean would be impossible without all sorts of different math applications. Even the mixtures of food we eat require an explanation based on math principles. In short, you might be surprised at just how many different ways MATLAB is being used, and this chapter tells you only about ten of the most popular uses. Explaining every potential MATLAB use would likely require an entire book.

Engineering New Solutions

Many engineering disciplines are out there, some of which are quite exotic while others are mundane. However, they all have one thing in common: They rely on various kinds of math to ensure that the results of any design process or new theory of how the universe works actually make sense. A new building isn't much use if it can't hold up to the stresses placed on it. Sending someone to Mars without doing the math work first could send that person to anywhere but Mars.

Creating new technologies means first computing how this new technology will work. Many ideas and concepts used in engineering rely on math as a starting point because math provides the means to express the idea in a form that others can understand. In short, if you're an engineer of any sort, you eventually need to perform math-related tasks that could benefit from MATLAB. You can get some additional ideas of just how useful MATLAB could be to you as an engineer by reading the discussion at `http://www.mathworks.com/matlabcentral/answers/72325-will-i-even-use-matlab-in-my-engineering-career`.

Getting an Education

Mathworks places a strong emphasis on education. In fact, you can find a special place for education-related materials at `http://www.mathworks.com/academia/`. The point of the education process is to obtain knowledge, of course, but it also means building certain disciplines (skills, methods of viewing information, and so on) — good habits is another way to look at it. To build these habits, you really need to have a product you can use that will allow you to practice performing tasks in the same way that you do after you leave school. Even if the organization that employs you doesn't use MATLAB, the principles you learn by working through problems with MATLAB follow standards that apply equally well to other products.

One of the more useful resources that you find on the MATLAB site at `http://www.mathworks.com/academia/student_center/tutorials/` is the list of tutorials. The tutorials help you get started using MATLAB and then take you through various kinds of tasks. After you finish reading this book, you can use these tutorials to increase your knowledge, take your education in a specific direction, and build the skills needed to either start or continue performing the work you do.

If you're a student and you need a copy of MATLAB, you can purchase it at `http://www.mathworks.com/academia/student_version/`. Several versions of MATLAB are available for student use, so make sure that you pick the package that best suits your needs. Note that you may require some special add-ons for your classes, which you can get at `https://www.mathworks.com/store/link/products/student/`.

Working with Linear Algebra

It may be hard to believe, but linear algebra really is part of the workplace (and not just for sending someone to the moon). For example, to calculate Return on Investment (ROI), you must know algebra. The same holds true for the following:

- ✔ Predicting the amount of turnover a company will have
- ✔ Determining how many items to keep in inventory
- ✔ Making life and business decisions, such as whether it's cheaper to rent a car or to buy one outright
- ✔ Creating a financial plan, such a determining whether it makes more sense to pay down a credit card or build up savings

No one would obtain MATLAB to perform these tasks just one time. However, if your job is helping people perform these sorts of tasks, you really do need something like MATLAB so that you can get the answers you need fast.

Uses for linear algebra often appear in places that you might never consider. For example, if you're a restaurant owner, you might use linear algebra on a regular basis to make your business more efficient. Check out the article at `http://smallbusiness.chron.com/restaurants-use-linear-programming-menu-planning-37132.html` for details. Imagine how surprised you might be if you walked into the back room of a restaurant sometime to find the manager poking away at a keyboard with a copy of MATLAB running!

Performing Numerical Analysis

Numerical analysis relies on approximation rather than the precision you see in symbolic math. It seems that the world is filled with approximations, and so is the galaxy. Performing certain building construction tasks is impossible without applying numerical analysis, and astronomy seems to require heavy use of it as well. You probably won't see a carpenter applying numerical analysis on the job site with MATLAB, but you will see architects who might need to do so.

Numerical analysis truly does apply to the natural world. For example, much of modern biology and medicine rest on principles described using numerical analysis. Your family physician probably doesn't require a copy of MATLAB, but the researcher who provides your physician with the information needed to diagnose any problems with your health does. When it comes to numeric analysis, you're better off thinking about the creative end of things rather than the application. The person inventing a new procedure needs MATLAB, but the person applying it on the job site doesn't.

Getting Involved in Science

Science is a pretty broad term, but it does have specific applications. MATLAB is likely to be used to explore new theories. It's important to differentiate between science and engineering in this case. *Engineering* is the application of known principles and theories to a problem in a predictable and usually standardized manner. *Science,* on the other hand, is the act of creating, testing, and proving principles and theories to eventually use to solve problems. In other words, when applied to science, MATLAB helps you perform "what if" analysis that helps you confirm (or deny) the viability of a theory.

Of course, science is used in many different ways. For example, you might be involved in the health industry and using science to find a cure for cancer or the Ebola virus. A computer scientist might look for a new way to use computer technology to aid those with accessibility needs. In fact, there are all sorts of ways in which MATLAB could figure into helping someone do something special for humanity.

Engaging Mathematics

MATLAB is all about working with mathematics in a comfortable environment that is less prone to errors. The rest of the chapter (in fact, most of this book) tells you all about this way of looking at mathematics. However, some people simply enjoy playing with math. It's the reason that so many theorems are available today to solve problems. These people are engaged with math in a way that few others can readily understand. MATLAB makes it possible to play with math, to create new ways of using numbers to perform useful tasks.

Exploring Research

Researchers have the world's best job in many respects. As a researcher, you get to ask a question, no matter how absurd, and determine whether the question is both answerable and relevant. After the question is answered, a researcher needs to determine whether the answer is both useful and reliable. In short, some view research as a kind of play (and they are correct — it really is play for the creative and intelligent mind).

Of course, research isn't just fun and games. If it were, people would have flocked to research as they do to video games now. After a question is asked and an answer is given, the researcher must convince colleagues that the answer is correct and then viable to put into practice. MATLAB lets you check the answer and verify that it does, in fact, work as the researcher suggests. After an answer is proven, the researcher can use MATLAB further to define precisely how the answer is used.

Although many of MATLAB's uses require graphics, research has a significant need for graphics because the researcher must often explain answers to people who don't have the researcher's skills. In most cases, the explanation will never work with a text-only presentation; the researcher must also include plenty of graphics that start with abstract concepts, and then turn them into concrete ideas that the audience can understand.

Walking through a Simulation

Flying a real jet is both dangerous and expensive — flying a ship carrying humans to Mars is impossible at the moment. *Simulation* is the art of determining whether something is possible, at least in theory, by using known facts. Using a simulation rather than a real-world counterpart is a low-cost approach to testing that is an essential part of any sort of scientific or engineering endeavor today, for these reasons:

- Saves human lives by seeing what could go wrong in the real-world environment

- Saves time because simulations are usually easier to set up

- Enhances the ability of the people involved to try various solutions

- Reduces costs by using and wasting fewer resources

- Improves the chances of a new technology succeeding by removing reliability issues before the real technology is built

- Increases the security surrounding a new technology by making it possible to test the technology without actually building a prototype that could be compromised

MATLAB makes simulations possible in several different ways. It may not always provide a complete solution, but you can use it to perform these kinds of tasks:

- Define the original math model used to define the technology and therefore the simulation

- Create individual snapshots showing how the technology will work based on the model

- Demonstrate the workflow for a technology using animation techniques so that even less-skilled stakeholders can see the technology at work

Employing Image Processing

Images are made up of pixels. Each *pixel* defines a particular color in a specific location in the image. In short, a pixel is a dot of just one color. The color is actually a numeric value that defines how much red, blue, and green to use in order to create the pixel color. Because the pixel is represented as a number, you can use various math techniques to manage the pixel. In fact, images are often managed as matrices, as you saw in several areas of the book (most notably Chapters 6, 7, and 16). A matrix is simply a structure consisting of numeric information.

Image processing is the act of managing the pixels in an image using math techniques to modify the matrix values. Techniques such as adding two matrices together are common when performing image processing. In fact, probably any technique you can think of that applies to matrices is also employed in image processing in some way.

The one thing you should know by this point in the book is that MATLAB excels at matrix manipulation. Anyone involved in image processing needs the sort of help that MATLAB provides to create and test new image processing techniques. The point is that you can test the math and then see the result right onscreen without changing applications. You can use MATLAB to both create the required math and then test it (at least in a simulated environment).

Embracing Programming Using Computer Science

The effect on the world at large with computer science is immense because the applications that programmers build affect every other scientific and engineering endeavor there is. In fact, the influence is even larger than that. Computer science defines how appliances work, how games play, and even how your doctor checks your health. In all these myriad uses, computer scientists rely heavily on math to perform tasks. MATLAB, with its rich toolbox, can be used to rapidly prototype an algorithm before committing the development resources to implementing the algorithm in another language, such as C++ or Java. Programmers commonly depend on MATLAB to enhance their productivity.

To give you some idea of just how much math is involved, consider the fact that you can't even write information to the screen without using math in some way. The position of the text must be calculated, as does the text size so that the computer knows whether the text will actually fit. All the fancy transitions you rely on to display special effects onscreen have their basis in math. In fact, computers have special components devoted solely to the pursuit of math because you can't do anything on a computer without it.

Of course, the question isn't whether computer scientists use math quite a lot — it's whether they can use MATLAB to do it. The answer is a resounding "yes." When creating an application, you must ensure that the output is valid. Otherwise, someone using the application will encounter errors (also called bugs) that will cause both the user and the computer scientist woe.

Verification is just one use. Someone who spends a lot of time working on computers developing applications needs to explore new ways of accomplishing tasks continually. Nothing is completely set with computers because there is always more than one way to accomplish any given task. In fact, often there isn't even a best way to perform the task, just a best way in a given circumstance. For a branch of science that should be closer to engineering than science, computer scientists deal with a nearly inexhaustible supply of unknowns that could potentially benefit from the use of MATLAB.

Chapter 18

Ten Ways to Make a Living Using MATLAB

*M*ATLAB is an excellent tool that performs a great many tasks. However, you might perceive it as just a tool, not a means by which to earn a living. Be ready to be amazed! Adding MATLAB proficiency to your résumé may be the thing that gets your foot in the door for that bigger, better job you've been wanting. In other words, MATLAB *is* the job, rather than a tool you use to perform the job, in at least some cases. Employers have learned to ask for MATLAB by name.

A look at Monster.com (http://jobsearch.monster.com/search/?q=matlab) clearly shows that proficiency in using MATLAB is the main educational requirement for getting some jobs. At the time of this writing, Monster.com had 766 postings specifically stating a need for MATLAB proficiency. When looking for work, don't overlook user groups such as the one on LinkedIn (http://www.linkedin.com/groups/MATLAB-Users-109866). These online sources often provide the networking needed to get just the right job. The following sections describe some of the more interesting ways (and just the tip of the iceberg at that) to get a job with your newfound MATLAB skills.

Working with Green Technology

At one point, energy came from one source — a power company. The grid could be simple because the energy went directly from one producer to multiple homes and businesses. Yes, there were interconnected grids, but the connections were relatively simple.

Today the energy picture is far different because many homes and businesses now produce energy in addition to using it. A solar panel on a home might produce enough energy to meet the home's needs and the needs of a neighboring home, which means that the grid has to be able to deal with a surplus from a home at times. In addition, the power company sources may now include wind farms in diverse locations.

The old grid system doesn't have the intelligence to manage such diverse power sources and sinks, so new smart grids are being used to replace the old system and make the new system more robust and flexible. Developing and implementing smart grids is a lot more difficult than working with the grids of old — MATLAB can help you perform the math required to make these new grids happen. You can read more about this exciting job at http://www.modelit.nl/index.php/matlab-webserver-user-story.

Looking for Unexploded Ordinance

You can't find an activity that's much more exciting than looking for unexploded ordinance. Let's just say that a cool resolve to get the job done and steady nerves are prerequisites for this job. However, most people don't understand the scope of the job. It's not just a few pieces of unexploded ordinance — it's thousands of pieces spread over a large area. Trying to find every last piece (and let's face it, when working with unexploded ordinance, even one piece left behind is one piece too many) is almost impossible without help.

People whose job is to locate unexploded ordinance use MATLAB to improve their chances of finding every last piece of it while keeping costs low. Most of these efforts have limited funds, so the challenge is to find more ordinance before the money runs out. Check out the article at http://www.mathworks.com/company/user_stories/Black-Tusk-Geophysics-Detects-and-Classifies-Unexploded-Ordnance.html to get additional information.

Creating Speech Recognition Software

The need for computers to recognize what humans are saying is increasing as humans rely more heavily on computers for help. Although products like Siri (`https://www.apple.com/ios/siri/`) are helpful, the uses for speech recognition go well beyond the basic need of turning human speech into something a computer can understand. Consider the fact that robots are becoming part of daily life. In fact, robots may eventually provide the means for people to stay in their homes at a time when they'd normally move to a nursing home (`http://whatsnext.blogs.cnn.com/2013/07/19/robots-the-future-of-elder-care/`). These robots will need to be able to recognize everything a person is saying and react to it. That's why ongoing research is important (`http://www.cs.dartmouth.edu/~dwagn/aiproj/speech.html`).

Expect to find a lot of jobs that work in the areas of speech in the future. Most of these jobs will require the ability to use math software, such as MATLAB, to speed up the process. Efficient use of math to help solve speech recognition problems will become more essential as the market for various types of computer-assisted technology increases.

Getting Disease under Control

Eliminating every disease may not be possible. Despite talk of a cure for the common cold, many other human ailments take precedence. MATLAB has an important role to play in the never-ending quest for the cure and management of disease. For example, the Centers for Disease Control (CDC) uses MATLAB for poliovirus sequencing and tracking. (See the story at `http://www.mathworks.com/company/user_stories/Centers-for-Disease-Control-and-Prevention-Automates-Poliovirus-Sequencing-and-Tracking.html`.)

The clock is ticking, and health organizations need every second they can get to chase down and kill off virulent diseases. Lest you think that modern medicine has been hugely successful in eliminating such diseases, think again. Only one serious pathogen has been eradicated to date: smallpox. (See the story at `http://www.historyofvaccines.org/content/articles/disease-eradication`.) We have a lot of work to do. Using products such as MATLAB makes researchers more efficient, creating the possibility of eradicating a virus or bacteria sooner — but only if the researcher actually knows how to use the software.

Becoming a Computer Chip Designer

Knowing how computers work at a detailed level — all the way down into the chip — opens an entirely new world of math calculations. A truly in-depth knowledge of computer chip technology involves not just electronics or chemistry, but a mix of both, with other technologies added in for good measure. (Strong math skills are a requirement because you can't actually see the interactions take place; you must know that they will occur based on the math involved.) The people who design chips today enter an alternative reality because things really don't work the way you think they will when you're working at the level of individual atoms. Yet, jobs exist for designing a System on a Chip (SoC) or Application Specific Integrated Circuit (ASIC). You can read about one such job at `http://jobview.monster.com/Sr-Systems-Engineer-SoC-ASIC-%E2%80%93-Matlab-Job-Beaverton-OR-137025672.aspx`. And given the rapid advances in chip technology, you can count on lots of opportunities in this area.

Keeping the Trucks Rolling

Designing better trucks may not seem like something you can do with MATLAB, but modern trucks are actually complicated machinery that has to operate safely on increasingly crowded roads. For example, just designing the air suspension systems used to couple the truck to the trailer and ensure that the two remain attached is a difficult task. You can read about this particular need at `http://www.mathworks.com/company/user_stories/Continental-Develops-Electronically-Controlled-Air-Suspension-for-Heavy-Duty-Trucks.html`. The point is that designing trucks is a complex task.

You can easily extend this particular need to other sorts of vehicles. The amount of engineering to design a modern car is daunting. Think about everything that a car needs to do now — everything from braking so that the car doesn't slide to ensuring that the people in the vehicle remain safe during an accident. Modern vehicles do all sorts of nonvehicle things such as entertain the kids in the back seat so that you can drive in peace. All these capabilities require engineering that is better done using math software.

Creating the Next Generation of Products

Research and Development (R&D) may not sound all that exciting at first. It really does need a better name because what you're really doing is creating all the products that people will use in the future. When working in R&D, an engineer or scientist gets to work with bleeding-edge technology — things that no one else can even imagine at times. You can read about one such job at `http://jobview.monster.com/Hands-On-Software-Developer-C-C-Matlab-Job-Colorado-Springs-CO-137454808.aspx`. However, given the amount of innovation going on, you can be sure that many other jobs exist in this area. Most of these jobs require proficiency using math software such as MATLAB. Time is money in these places, and they want the people working on tomorrow's technology to be as efficient as possible.

Designing Equipment Used in the Field

Even though the story at `http://www.mathworks.com/company/user_stories/Electrodynamics-Associates-Designs-High-Performance-Generator-Controller-for-the-Military.html` tells about a generator designed for use by the military, the story speaks to a much larger need. Every outdoor activity, such as construction projects, requires equipment of various sorts to make the task doable. Generating power is a huge need. Trying to build something would be nearly impossible without the electricity required to run various kinds of tools.

Generators, tools, appliances, and all sorts of other devices require special design today. Not only does the equipment need to work well, but it has to do so at a low energy cost, reduced use of materials, nearly invisible mainte-nance costs, and with a small environmental impact. Trying to design such equipment without the proper math software would be nearly impossible.

Performing Family Planning

Researchers are currently using products such as MATLAB to help predict how the use of contraceptives and other family-planning methods will help reduce populations in places like India and Bangladesh. Trying to find tech-niques that work well to help control populations so that the local economy can support them is hard — and ensuring that those techniques are actually working is harder still. This is the sort of job that requires a mix of human and research skills. You'd use the biological add-ins for MATLAB in this case. One such instance of advanced techniques at work is at `http://econ.worldbank.org/external/default/main?pagePK=64165259&theSitePK=477894&entityID=000009265_3961003174232`.

Reducing Risks Using Simulation

Many activities entail risk. It isn't a matter of throwing money at the problem until you solve it; you really can't know that the project will succeed at all. Using simulation can greatly increase the likelihood of success, however. For example, consider the salvaging of the Russian submarine Kursk. (See the story at `http://www.mathworks.com/company/user_stories/International-Salvage-Team-Brings-Home-the-Kursk-Submarine-Using-a-Simulation-Developed-in-Simulink.html` for details.) It wouldn't be possible to know whether the salvage could ever succeed without simulating it first. Simulations require both math skills and plotting, both of which are found in MATLAB.

However, just knowing that the project can succeed isn't enough. Using a simulation lets you identify potential risks at the outset, before the project is under way. Risk identification and management are important parts of many endeavors today. Beyond having procedures in place that guarantee success, you need procedures for those times when things do go wrong. Creating a great simulation helps an engineer with the required knowledge to figure out the risky situations in advance and do everything needed to avoid them, and then also create procedures for when things go wrong anyway.

Appendix A

MATLAB Functions

• •

*T*his appendix provides you with an overview of the MATLAB functions. Each function helps you perform a specific task within MATLAB. The tables contain the function name and a short description of the task that the function performs. If you need more information than this appendix gives you, you can obtain additional information by typing **help** *<function_name>* in the MATLAB command window and pressing Enter. Of course, you also have the information found in this book, and you can search in Help. See Chapter 3 for some useful information on using the Help resources that MATLAB provides.

Table A-1	Arithmetic Functions
Function	*Description*
uplus	Unary plus — comparable to + acting on one object
plus	Plus — comparable to + adding two objects
uminus	Unary minus — comparable to – acting on one object
minus	Minus — comparable to – subtracting two objects
mtimes	Matrix multiplication — comparable to *
times	Array multiply — comparable to .*
rdivide	Right-array division — comparable to ./
mrdivide	Solves systems of linear equations xA = B for x — comparable to /
ldivide	Left-array division — comparable to .\
mldivide	Solves systems of linear equations Ax = B for x — comparable to \
power	Array power — comparable to .^
mpower	Matrix power — comparable to ^
diff	Differences and approximate derivatives
prod	Product of array elements

(continued)

Table A-1 *(continued)*

Function	Description
sum	Sum of array elements
ceil	Rounds toward positive infinity
fix	Rounds toward zero
floor	Rounds toward negative infinity
idivide	Integer division with rounding option
mod	Modulus after division
rem	Remainder after division
round	Rounds to nearest integer

Table A-2	Trigonometric Functions
Function	**Description**
sin	Sine of argument in radians
sind	Sine of argument in degrees
asin	Inverse sine; result in radians
asind	Inverse sine; result in degrees
sinh	Hyperbolic sine of argument in radians
asinh	Inverse hyperbolic sine
cos	Cosine of argument in radians
cosd	Cosine of argument in degrees
acos	Inverse cosine; result in radians
acosd	Inverse cosine; result in degrees
cosh	Hyperbolic cosine
acosh	Inverse hyperbolic cosine
tan	Tangent of argument in radians
tand	Tangent of argument in degrees
atan	Inverse tangent; result in radians
atand	Inverse tangent; result in degrees
atan2	Four-quadrant inverse tangent
atan2d	Four-quadrant inverse tangent; result in degrees
tanh	Hyperbolic tangent

Function	Description
atanh	Inverse hyperbolic tangent
csc	Cosecant of argument in radians
cscd	Cosecant of argument in degrees
acsc	Inverse cosecant; result in radians
acscd	Inverse cosecant; result in degrees
csch	Hyperbolic cosecant
acsch	Inverse hyperbolic cosecant
sec	Secant of argument in radians
secd	Secant of argument in degrees
asec	Inverse secant; result in radians
asecd	Inverse secant; result in degrees
sech	Hyperbolic secant
asech	Inverse hyperbolic secant
cot	Cotangent of argument in radians
cotd	Cotangent of argument in degrees
acot	Inverse cotangent; result in radians
acotd	Inverse cotangent; result in degrees
coth	Hyperbolic cotangent
acoth	Inverse hyperbolic cotangent
hypot	Square root of sum of squares

Table A-3	Exponentials, Logarithms, Powers, and Roots
Function	**Description**
exp	Exponential
expm1	Computes exp(x)–1 accurately for small values of x
log	Natural logarithm
log10	Common (base 10) logarithm
log1p	Computes log(1+x) accurately for small values of x
log2	Determines base 2 logarithm and dissect floating-point numbers into exponent and mantissa
nthroot	Real nth root of real numbers
pow2	Base 2 power and scale floating-point numbers

(continued)

Table A-3 *(continued)*

Function	Description
reallog	Natural logarithm for nonnegative real arrays
realpow	Array power for real-only output
realsqrt	Square root for nonnegative real arrays
sqrt	Square root

Table A-4   Complex Number Functions

Function	Description
abs	Absolute value and complex magnitude
angle	Phase angle
complex	Construct complex data from real and imaginary components
conj	Complex conjugate
i	Imaginary unit
imag	Imaginary part of complex number
isreal	Checks whether input is a real array
j	Imaginary unit
real	Real part of complex number
sign	Signum() function, which returns 1 if the corresponding element is greater than 0, 0 if the corresponding element is zero, and −1 if the corresponding element is less than 0.

Table A-5   Discrete Math Functions

Function	Description
factor	Prime factors
factorial	Factorial function
gcd	Greatest common divisor
isprime	Array elements that are prime numbers
lcm	Least common multiple
nchoosek	Binomial coefficient or all combinations
perms	All possible permutations
primes	Generates list of prime numbers
rat, rats	Rational fraction approximation

Table A-6	Polynomial Functions
Function	**Description**
poly	Polynomial with specified roots
polyder	Polynomial derivative
polyeig	Polynomial eigenvalue problem
polyfit	Polynomial curve fitting
polyint	Integrates the polynomial analytically
polyval	Polynomial evaluation
roots	Polynomial roots

Table A-7	Special Functions
Function	**Description**
erf	Error function
erfc	Complementary error function
erfcinv	Inverse complementary error function
erfcx	Scaled complementary error function
erfinv	Inverse error function

Table A-8	Cartesian, Polar, and Spherical Functions
Function	**Description**
cart2pol	Transforms Cartesian coordinates to polar or cylindrical coordinates
cart2sph	Transforms Cartesian coordinates to spherical coordinates
pol2cart	Transforms polar or cylindrical coordinates to Cartesian coordinates
sph2cart	Transforms spherical coordinates to Cartesian coordinates

Table A-9	Constants and Test Matrix Functions
Function	**Description**
eps	Floating-point relative accuracy
Inf	Infinity

(continued)

Table A-9 *(continued)*

Function	Description
pi	Ratio of circle's circumference to its diameter
NaN	Not-a-Number
isfinite	Array elements that are finite
isinf	Array elements that are infinite
isnan	Array elements that are NaN
gallery	Test matrices
magic	Magic square

Table A-10   **Matrix Operation Functions**

Function	Description
cross	Vectors cross product
dot	Vectors dot product
kron	Kronecker tensor product
transpose	Transposes

Table A-11   **Linear Equation Functions**

Function	Description
inv	Matrix inverse
linsolve	Solves linear system of equations

Table A-12   **Eigenvalue Functions**

Function	Description
eig	Eigenvalues and eigenvectors
eigs	Largest eigenvalues and eigenvectors of matrix
sqrtm	Matrix square root

Table A-13	Matrix Analysis Functions
Function	*Description*
det	Matrix determinant
norm	Vector and matrix norms
rank	Rank of matrix
rref	Reduced row echelon form
trace	Sum of diagonal elements

Table A-14	Matrix Functions
Function	*Description*
expm	Matrix exponential
logm	Matrix logarithm
sqrtm	Matrix square root
arrayfun	Applies function to each element of array

Table A-15	Statistical Functions
Function	*Description*
corrcoef	Correlation coefficients
cov	Covariance matrix
max	Largest elements in array
mean	Average or mean value of array
median	Median value of array
min	Smallest elements in array
mode	Most frequent values in array
std	Standard deviation
var	Variance

Table A-16	Random Number Generator
Function	*Description*
rng	Controls random number generation
rand	Uniformly distributed pseudo-random numbers
randn	Normally distributed pseudo-random numbers
randi	Uniformly distributed pseudo-random integers

Table A-17	1-D Interpolation
Function	*Description*
interp1	1-D data interpolation (table lookup)
Spline	Cubic spline data interpolation

Table A-18	Gridded Data Interpolation
Function	*Description*
interp2	2-D data interpolation (table lookup)
interp3	3-D data interpolation (table lookup)
interpn	nD data interpolation (table lookup)
griddedInterpolant	Interpolant for gridded data
ndgrid	Rectangular grid in nD space
meshgrid	Rectangular grid in 2D and 3D space

Table A-19	Scattered Data Interpolation
Function	*Description*
griddata	Interpolates scattered data

Table A-20	Optimization Functions
Function	*Description*
fminbnd	Finds minimum of single-variable function on fixed intervals
fminsearch	Finds minimum of unconstrained multivariable function using derivative-free method
fzero	Finds root of continuous function of one variable

Table A-21	Ordinary Differential Equation Functions
Function	*Description*
ode23	Solve nonstiff differential equations; low-order method

Table A-22	Sparse Matrix Manipulation Functions
Function	*Description*
spy	Visualizes a sparsity pattern
find	Finds indices and values of nonzero elements

Table A-23	Elementary Polygons
Function	*Description*
polyarea	Area of polygon
inpolygon	Points inside polygonal region
rectint	Rectangle intersection area

Appendix B

MATLAB's Plotting Routines

● ●

*T*his appendix lists the plotting routines in MATLAB with a brief description and example. To save space, some entries show a single encompassing example for multiple commands. An example may also use an ellipsis (...) to show the continuation of a pattern. To assist you, we have included several functions to generate test matrices, for example, `rand()`, `magic()`, `peaks()`, `cylinder()`, `ellipsoid()`, and `sphere()`.

Table B-1	Basic Plotting Routines	
Routine	*Description*	*Example*
ezplot	Plots the expression enclosed in quotes	`ezplot('exp(-0.4*x)*sin(x)')`
fplot	Same as `ezplot`, but requires limits	`fplot('exp(-0.4*x)*sin(x)', [0,2*pi,-0.3,0.6])`
plot	Plots data passed in by vectors	`x=[0:2*pi/100:2*pi];` `y=exp(-0.4*x).*sin(x);`
comet	Just like `plot`, but `comet` animates the trajectory. It helps to have a larger vector to slow down the comet trace a tad	`plot(x,y); figure(2)` `comet(x,y); figure(3)` `ribbon(x,y);figure(4)` `y2=100*exp(-0.4*x).*cos(x);` `plotyy(x,y,x,y2)`
ribbon	Like `plot`, but it displays the data as 3D ribbons	
plotyy	Plots data where y values may differ greatly — it makes two y-axes	

(continued)

Table B-1 *(continued)*

Routine	Description	Example

loglog	Plots data on a log scale on both x- and y-axes — y proportional to a power of x is straight on this plot	`x=[0:2*pi/100:2*pi];`
		`y=10*x.^pi;`
		
		`loglog(x,y)`

semilogx	X-axis data is on a log scale and the y-axis is on a linear scale — y linearly related to log(x) is straight on this plot	`x=[0:2*pi/100:2*pi];`
		`y=10*log(x)+pi;`
		
		`semilogx(x,y);`

semilogy	Y-axis data is on a log scale and x-axis linear — y proportional to exponential of x is straight on this plot	`x=[0:2*pi/100:2*pi];`
		`y=10*exp(pi*x);`
		
		`semilogy(x,y);`

Table B-2   Beyond Basics

Routine	Description	Example

area	Acts just like `plot` except for the fact that it fills in the area for you	`x=[0:2*pi/100:2*pi];`
		`y=exp(-0.4*x).*sin(x);`
		`area(x,y)`

pie	Creates a standard pie chart	`x=[2,4,6,8];` `pie(x); figure(2)`
pie3	Adds some 3D pizazz to `pie`	`pie3(x)`

Routine	Description	Example
ezpolar	Creates a polar plot where the distance from origin vs. angle is plotted — argument is function expression	`ezpolar('cos(2*x)^2');`

polar	Creates a polar plot like ezpolar but does so by accepting vector arguments — x value corresponds to angle and y value to distance from origin	`x=[0:2*pi/100:2*pi];`
		`y=(exp(-0.1*x).*sin(x)).^2;`
		`polar(x,y)`

compass	Like polar, but compass shows data as vectors from origin	`compass(rand(1,3)-0.5,...` `rand(1,3)-0.5)`

bar	Creates a standard bar chart that can handle both grouping and stacking	`x=[8,7,6;13,21,15;32,27,32];`
		`bar(x); figure(2)`
		`y=sum(x,2);`

barh	Just like bar except plot is horizontal	`bar(y); figure(3)` `bar(x,'stacked'); figure(4)`
bar3	Adds a little 3D pizazz to a standard bar chart	`barh(x); figure(5)` `bar3(x); figure(6)`
bar3h	Creates a 3D horizontal bar chart	`bar3h(x)`

fill	Fills polygons (vector inputs define vertices) with the specified color	`y=sin([0:2*pi/5:2*pi])`
		`x=cos([0:2*pi/5:2*pi])`
		`fill(x,y,'g')`

Table B-3	Statistical Plotting Routines	
Routine	*Description*	*Example*
`scatter`	Plots `(x,y)` data points.	`x=[0:2*pi/100:2*pi];` `y=10*x+pi+10*randn(1,101);`
`stem`	Like `scatter`, but adds a line from the x axis to a data point.	`scatter(x,y); figure(2)` `stem(x,y); figure(3)`
`errorbar`	Like `scatter` but adds error bars.	`errorbar(x,y,10*ones(1,` `101),ones(1,101))` `figure(4);`
`hist` or `histogram`	Creates a histogram — a bar chart showing the frequency of occurrence of data vs. value. These two routines handle the bins differently.	`hist(y); figure(5)` `histogram(y)` `histc(y,[-40:20:80])` `histcounts(y,5)`
`histc`	Related to `hist`, but rather than making a plot, it makes a vector of counts.	`histc(y,[-40:20:80])` example with `histc` category
`histcounts`	Like `histc` except it creates *n* bins.	`histcounts(y,5)` example with `histcounts` category
`stairs`	Like `scatter`, but makes stairsteps when y values change.	`x=[0:2*pi/10:2*pi];` `y=10*x+pi+10*randn(1,11);` `stairs(x,y)`
`rose`	A cross between `polar` and `histogram`; it displays frequency vs. angle.	`rose(randn(1,100),5)`
`pareto`	A bar chart arranged with the highest bars first.	`histc(randn(1,100),` `[-4:1:4])` `pareto(ans)`

Routine	Description	Example

spy	A scatter plot of zeros in a matrix.	`mymat=rand(5);`
		`mymat=(mymat>0.5).*mymat;`
		
		`spy(mymat)`

plotmatrix	A scatter plot of all permutations of columns of x and y.	`plotmatrix(magic(3),` `magic(3))`

Table B-4		**3D Graphics**	

Routine	Description	Example
ezcontour	Makes a contour plot like a topographic map	`ezcontour('cos(x)*` `cos(y)')`
ezcontourf	Same as ezcontour except it fills in the area between contours	`ezcontourf('cos(x)*` `cos(y)')`
ezmesh	Creates a 3D perspective plot with an open wireframe mesh	`ezmesh('cos(x)*` `cos(y)')`
ezsurf	Creates a 3D perspective plot with a filled surface	`ezsurf('cos(x)*` `cos(y)')`
ezmeshc	Combines a contour plot with mesh	`ezmeshc('cos(x)*` `cos(y)')`
ezsurfc	Combines a contour plot with a 3D perspective filled surface plot	`ezsurfc('cos(x)*` `cos(y)')`

ezplot3	Plots a curve in 3D space; capable of being animated	`ezplot3('sin(x)',`
		`'cos(x)','sin(3*x/2)',...`
		
		`[-2*pi,2*pi],'animate')`

plot3	Like ezplot3, except it takes vector arguments	`a=[-2*pi:4*pi/100:2*pi];`
		
		`x=sin(a); y=cos(a);`
		`z=sin(3*a/2);`

scatter3	Like plot3, but shows individual points	`plot3(x,y,z);`
		`figure(2)`
		
		`scatter3(x,y,z)`

(continued)

Table B-4 *(continued)*

Routine	Description	Example
stem3	Stem plot of 3D data	stem3(rand(5))
contour	Creates a contour plot with matrix arguments	x=[-2*pi:4*pi/100:2*pi];
contourf	Same as contour, except contourf fills in the contours	y=[-2*pi:4*pi/100:2*pi]; z=cos(x)'*cos(y);
contour3	Same as contour, except contour3 provides a 3D perspective	contour(x,y,z);title ('contour');figure(2)
		contourf(x,y,z);title ('contourf');figure(3)
surf or surface	Creates a filled surface	contour3(x,y,z);title ('contour3');figure(4)
mesh	Creates a wireframe mesh surface	surf(x,y,z);title ('surf');figure(5)
waterfall	Like mesh, but all column lines are omitted	surface(x,y,z);title ('surface');figure(6)
surfc	Same as surf, but with contour plot added	mesh(x,y,z);title ('mesh');figure(7)
meshc	Wireframe mesh with contour	waterfall(x,y,z); title('waterfall'); figure(8)
meshz	Wireframe mesh with a curtain around the plot	surfc(x,y,z);title ('surfc');figure(9)
surfl	Same as surf, except surfl simulates light and shadow	meshc(x,y,z);title ('meshc');figure(10)
		meshz(x,y,z);title ('meshz');figure(11)
pcolor	Shows values of the matrix as colors	surfl(x,y,z);title ('surfl');figure(12)
		pcolor(z);title('surfl')
surfnorm	Creates surface plot with normal vectors	[x,y,z]=peaks; % Test function
		surfnorm(x,y,z)
fill3	Like fill except in 3D; note that the fill area may not be coplanar	fill3([0,1,1,0],[0,0, 1,0],[0,1,0,1],'g')

Table B-5		Vector Fields
Routine	*Description*	*Example*
`feather`	Like `compass` except it steps once for each element in `x` and `y`	`feather(rand(1,3)-0.5,...` `rand(1,3)-0.5)`
`quiver`	Works like `feather` except `quiver` plots vectors in an x-y plane	`[x,y]=meshgrid([-5:5],[-5:5]);` `u=ones(11)+(4./(sqrt(x.^2+y.^2)).` `*cos(atan2(y,x)));`
`stream line`	Plots line from the vector field; the example plots 2D `streamline` on the same plot as `quiver`	`v=(4./(sqrt(x.^2+y.^2)).*sin` `(atan2(y,x)));` `v(6,6)=0; u(6,6)=0;` `quiver(x,y,u,v); hold on` `streamline(x,y,u,v,[-5,-5,-5],` `[-1,.01,1]); hold off,` `figure(2)`
`quiver3`	Like `quiver` for 3D; the example adds a uniform velocity field to a $1/r^2$ velocity field	`[x,y,z]=meshg` `rid([-5:2:3],[-3:2:3],[-3:2:3]);` `r=sqrt(x.^2+y.^2+z.^2);` `u=ones(4,5,4)+(10./r.^2).*cos` `(atan2(y,x)).*sin(acos(z./r));`
`stream ribbon`	Like `streamline`, but shows ribbons	`v=(10./r.^2).*sin(atan2(y,x)).*` `sin(acos(z./r));`
`coneplot`	Like `quiver3`, but `coneplot` shows velocity as cones	`w=10.*z./r;` `quiver3(x,y,z,u,v,w);hold on;` `streamribbon(x,y,z,u,v,w,` `-5,0,.1);figure(2)`
`stream tube`	Like `streamline`, but plots tubes (cylindrical 3D flow lines)	`coneplot(x,y,z,u,v,w,x,y,z);` `figure(3)` `quiver3(x,y,z,u,v,w);hold on;` `streamtube(x,y,z,u,v,w,-5,0,.1);`

Index

About the Authors

Jim Sizemore is a physics and engineering professor earning an MS in physics from UC-San Diego and a PhD in Materials Science from Stanford University. He was employed many years in the semiconductor industry working on several projects, including diffusion and oxidation, radiation hardening, and optoelectronics. After his private sector career, he turned to teaching, and is currently a physics and engineering professor at Tyler Junior College in Tyler, Texas. There, he started a popular science club where students were able to design and build several projects, including a 2 meter trebuchet, just in case they needed to attack any castles in the area. He currently teaches programming for engineers, with MATLAB being the primary language taught. (Check out his instructional web site at `http://iteach.org/funphysicist/`.) He has enthusiasm for teaching and learning, but also enjoys photography and bicycle riding in his spare time.

John Mueller is a freelance author and technical editor. He has writing in his blood, having produced 96 books and more than 300 articles to date. The topics range from networking to artificial intelligence and from database management to heads-down programming. Some of his current books include a book on Python for beginners, a Java e-learning kit, a book on HTML5 development with JavaScript, and another on CSS3. His technical editing skills have helped over more than 63 authors refine the content of their manuscripts. John has provided technical editing services to both *Data Based Advisor* and *Coast Compute* magazines. It was during his time with Data Based Advisor that John was first exposed to MATLAB and he has continued to follow the progress in MATLAB development ever since. Be sure to read John's blog at `http://blog.johnmuellerbooks.com/`.

When John isn't working at the computer, you can find him outside in the garden, cutting wood, or generally enjoying nature. John also likes making wine, baking cookies, and knitting. When not occupied with anything else, he makes glycerin soap and candles, which come in handy for gift baskets. You can reach John on the Internet at John@JohnMuellerBooks.com. John is also setting up a website at `http://www.johnmuellerbooks.com/`. Feel free to take a look and make suggestions on how he can improve it.

Dedication

Jim Sizemore: I wish to dedicate this book to many mentors including Bob Abel, Greg Sherman, and Gene Branum. Also my thoughts and dedication go to my brother Bill who is afflicted with Parkinson's disease and to my son Daniel as the future belongs to his generation.

John Paul Mueller: This book is dedicated to Aaron and Sarah Oberman — two devoted people who delight in doing kind deeds for others in need.

Authors' Acknowledgments

Jim Sizemore: Thanks to Sara McCaslin who freely shared in order to start our MATLAB programming course.

Thanks also go to Gene Branum, Doug Parsons, and other Tyler Junior College colleagues who supported me in order to devote time to this project.

Finally, I wish to thank my coauthor John Mueller, whose writing skill and experience were essential to taking my original vision and finely polishing it.

John Paul Mueller: Thanks to my wife, Rebecca. Even though she is gone now, her spirit is in every book I write, in every word that appears on the page. She believed in me when no one else would.

Russ Mullen deserves thanks for his technical edit of this book. He greatly added to the accuracy and depth of the material you see here. Russ worked exceptionally hard helping with the research for this book by locating hard to find URLs and also offering a lot of suggestions.

Matt Wagner, my agent, deserves credit for helping me get the contract in the first place and taking care of all the details that most authors don't really consider. I always appreciate his assistance. It's good to know that someone wants to help.

A number of people read all or part of this book to help me refine the approach, test scripts, and generally provide input that all readers wish they could have. These unpaid volunteers helped in ways too numerous to mention here. I especially appreciate the efforts of Eva Beattie and Glenn A. Russell, who provided general input, read the entire book, and selflessly devoted themselves to this project.

Finally, I would like to thank Paul Levesque, Susan Christophersen, and the rest of the editorial and production staff.

Publisher's Acknowledgments

Acquisitions Editor: Connie Santisteban,
Andy Cummings

Senior Project Editor: Paul Levesque

Copy Editor: Susan Christopherson

Technical Editor: Russ Mullen

Editorial Assistant: Claire Johnson

Sr. Editorial Assistant: Cherie Case

Project Coordinator: Patrick Redmond

Cover Image: ©iStock.com/elly99

Apple & Mac

iPad For Dummies,
6th Edition
978-1-118-72306-7

iPhone For Dummies,
7th Edition
978-1-118-69083-3

Macs All-in-One
For Dummies, 4th Edition
978-1-118-82210-4

OS X Mavericks
For Dummies
978-1-118-69188-5

Blogging & Social Media

Facebook For Dummies,
5th Edition
978-1-118-63312-0

Social Media Engagement
For Dummies
978-1-118-53019-1

WordPress For Dummies,
6th Edition
978-1-118-79161-5

Business

Stock Investing
For Dummies, 4th Edition
978-1-118-37678-2

Investing For Dummies,
6th Edition
978-0-470-90545-6

Personal Finance
For Dummies, 7th Edition
978-1-118-11785-9

QuickBooks 2014
For Dummies
978-1-118-72005-9

Small Business Marketing
Kit For Dummies,
3rd Edition
978-1-118-31183-7

Careers

Job Interviews
For Dummies, 4th Edition
978-1-118-11290-8

Job Searching with Social
Media For Dummies,
2nd Edition
978-1-118-67856-5

Personal Branding
For Dummies
978-1-118-11792-7

Resumes For Dummies,
6th Edition
978-0-470-87361-8

Starting an Etsy Business
For Dummies, 2nd Edition
978-1-118-59024-9

Diet & Nutrition

Belly Fat Diet For Dummies
978-1-118-34585-6

Mediterranean Diet
For Dummies
978-1-118-71525-3

Nutrition For Dummies,
5th Edition
978-0-470-93231-5

Digital Photography

Digital SLR Photography
All-in-One For Dummies,
2nd Edition
978-1-118-59082-9

Digital SLR Video &
Filmmaking For Dummies
978-1-118-36598-4

Photoshop Elements 12
For Dummies
978-1-118-72714-0

Gardening

Herb Gardening
For Dummies, 2nd Edition
978-0-470-61778-6

Gardening with Free-Range
Chickens For Dummies
978-1-118-54754-0

Health

Boosting Your Immunity
For Dummies
978-1-118-40200-9

Diabetes For Dummies,
4th Edition
978-1-118-29447-5

Living Paleo For Dummies
978-1-118-29405-5

Big Data

Big Data For Dummies
978-1-118-50422-2

Data Visualization
For Dummies
978-1-118-50289-1

Hadoop For Dummies
978-1-118-60755-8

Language & Foreign Language

500 Spanish Verbs
For Dummies
978-1-118-02382-2

English Grammar
For Dummies, 2nd Edition
978-0-470-54664-2

French All-in-One
For Dummies
978-1-118-22815-9

German Essentials
For Dummies
978-1-118-18422-6

Italian For Dummies,
2nd Edition
978-1-118-00465-4

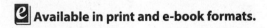 **Available in print and e-book formats.**

Math & Science

Algebra I For Dummies,
2nd Edition
978-0-470-55964-2

Anatomy and Physiology
For Dummies, 2nd Edition
978-0-470-92326-9

Astronomy For Dummies,
3rd Edition
978-1-118-37697-3

Biology For Dummies,
2nd Edition
978-0-470-59875-7

Chemistry For Dummies,
2nd Edition
978-1-118-00730-3

1001 Algebra II Practice
Problems For Dummies
978-1-118-44662-1

Microsoft Office

Excel 2013 For Dummies
978-1-118-51012-4

Office 2013 All-in-One
For Dummies
978-1-118-51636-2

PowerPoint 2013
For Dummies
978-1-118-50253-2

Word 2013 For Dummies
978-1-118-49123-2

Music

Blues Harmonica
For Dummies
978-1-118-25269-7

Guitar For Dummies,
3rd Edition
978-1-118-11554-1

iPod & iTunes
For Dummies, 10th Edition
978-1-118-50864-0

Programming

Beginning Programming
with C For Dummies
978-1-118-73763-7

Excel VBA Programming
For Dummies, 3rd Edition
978-1-118-49037-2

Java For Dummies,
6th Edition
978-1-118-40780-6

Religion & Inspiration

The Bible For Dummies
978-0-7645-5296-0

Buddhism For Dummies,
2nd Edition
978-1-118-02379-2

Catholicism For Dummies,
2nd Edition
978-1-118-07778-8

Self-Help & Relationships

Beating Sugar Addiction
For Dummies
978-1-118-54645-1

Meditation For Dummies,
3rd Edition
978-1-118-29144-3

Seniors

Laptops For Seniors
For Dummies, 3rd Edition
978-1-118-71105-7

Computers For Seniors
For Dummies, 3rd Edition
978-1-118-11553-4

iPad For Seniors
For Dummies, 6th Edition
978-1-118-72826-0

Social Security
For Dummies
978-1-118-20573-0

Smartphones & Tablets

Android Phones
For Dummies, 2nd Edition
978-1-118-72030-1

Nexus Tablets
For Dummies
978-1-118-77243-0

Samsung Galaxy S 4
For Dummies
978-1-118-64222-1

Samsung Galaxy Tabs
For Dummies
978-1-118-77294-2

Test Prep

ACT For Dummies,
5th Edition
978-1-118-01259-8

ASVAB For Dummies,
3rd Edition
978-0-470-63760-9

GRE For Dummies,
7th Edition
978-0-470-88921-3

Officer Candidate Tests
For Dummies
978-0-470-59876-4

Physician's Assistant Exam
For Dummies
978-1-118-11556-5

Series 7 Exam For Dummies
978-0-470-09932-2

Windows 8

Windows 8.1 All-in-One
For Dummies
978-1-118-82087-2

Windows 8.1 For Dummies
978-1-118-82121-3

Windows 8.1 For Dummies,
Book + DVD Bundle
978-1-118-82107-7

Available in print and e-book formats.

Available wherever books are sold. **For more information or to order direct visit www.dummies.com**